**Springer Theses**

Recognizing Outstanding Ph.D. Research

## Aims and Scope

The series "Springer Theses" brings together a selection of the very best Ph.D. theses from around the world and across the physical sciences. Nominated and endorsed by two recognized specialists, each published volume has been selected for its scientific excellence and the high impact of its contents for the pertinent field of research. For greater accessibility to non-specialists, the published versions include an extended introduction, as well as a foreword by the student's supervisor explaining the special relevance of the work for the field. As a whole, the series will provide a valuable resource both for newcomers to the research fields described, and for other scientists seeking detailed background information on special questions. Finally, it provides an accredited documentation of the valuable contributions made by today's younger generation of scientists.

## Theses are accepted into the series by invited nomination only and must fulfill all of the following criteria

- They must be written in good English.
- The topic should fall within the confines of Chemistry, Physics, Earth Sciences, Engineering and related interdisciplinary fields such as Materials, Nanoscience, Chemical Engineering, Complex Systems and Biophysics.
- The work reported in the thesis must represent a significant scientific advance.
- If the thesis includes previously published material, permission to reproduce this must be gained from the respective copyright holder.
- They must have been examined and passed during the 12 months prior to nomination.
- Each thesis should include a foreword by the supervisor outlining the significance of its content.
- The theses should have a clearly defined structure including an introduction accessible to scientists not expert in that particular field.

More information about this series at http://www.springer.com/series/8790

Shun Omagari

# Energy Transfer Processes in Polynuclear Lanthanide Complexes

Doctoral Thesis accepted by
Hokkaido University, Sapporo, Japan

Springer

*Author*
Dr. Shun Omagari
Graduate School of Chemical
Sciences and Engineering
Hokkaido University
Sapporo, Hokkaido, Japan

*Supervisor*
Prof. Dr. Yasuchika Hasegawa
Faculty of Engineering
Hokkaido University
Sapporo, Hokkaido, Japan

ISSN 2190-5053 ISSN 2190-5061 (electronic)
Springer Theses
ISBN 978-981-13-6048-0 ISBN 978-981-13-6049-7 (eBook)
https://doi.org/10.1007/978-981-13-6049-7

Library of Congress Control Number: 2018966863



This Springer imprint is published by the registered company Springer Nature Singapore Pte Ltd.
The registered company address is: 152 Beach Road, #21-01/04 Gateway East, Singapore 189721, Singapore

# Supervisor's Foreword

Lanthanide physics and chemistry have been studied ever since the discovery of rare earth elements in 1787 by C. A. Arrhenius, and extensive research on the photophysical properties of lanthanide compounds came later, after the ground-breaking theoretical works independently by B. R. Judd and G. S. Ofelt in 1962 (now known as the Judd–Ofelt theory). The high chromaticity and the long emission lifetimes of lanthanides are of particular importance in current technologies involving emission of light, ranging from commercialized technologies like lighting and telecommunication to research technologies like bioprobes and lasers. Combined with coordination chemistry, otherwise poorly light-emitting lanthanides can be enhanced into strongly light-emitting lanthanide complexes. The research field of lanthanide complexes is of great interest today with more than 1700 papers being published annually.

The quest for a Ph.D. degree in chemistry is daunting for the many because of the competitiveness, the requirement of full dedication, and the relatively long duration of the course. However, this did not prevent Shun Omagari from embarking on this quest, associated with the goal of unraveling the luminescence mechanism of polynuclear lanthanide complexes involving energy transfer. Specifically, the goal was about understanding the luminescence mechanism of lanthanide clusters, a polynuclear lanthanide complex with lanthanides "core" and ligands "shell."

Recently, polynuclear lanthanide complexes are considered as a promising class of luminescent material due to the energy transfer between lanthanide ions that give rise to new photofunctions such as temperature sensing, local probes, dual emission, and upconversion. Simultaneously, the additional processes lead to difficulties in analyzing the luminescence mechanism. Most reports on this topic usually involve a practical approach; they discuss the luminescence mechanism based on many experimental results and empirical evidence from other reports.

Shun took a different approach in his study, motivated by his fascination for physics and mathematics. He overcame the theoretical difficulties by carefully examining the Judd–Ofelt theory and kinetic analyses to reduce complicated problems into a simpler form and deliberately designed the experiments. The thesis first investigated the fundamental effect of energy transfer between $Yb^{3+}$ ions on the emission lifetimes of Yb coordination polymers. A qualitative (and quantitative with some assumptions) agreement was achieved between theoretical and experimental emission lifetimes when phonon-assisted energy transfer was taken into account.

The thesis also investigates the basic photophysical properties of nonanuclear lanthanide clusters. This includes the effect of spin–orbit coupling induced by the intrinsic angular momentum of electrons in $Gd^{3+}$, $Lu^{3+}$, and $Yb^{3+}$ ions on the energy transfer efficiency from the organic ligands to $Yb^{3+}$ ions. The result was that a spin–orbit coupling too strong could reduce the energy transfer efficiency, contrary to the convention that the stronger the spin–orbit coupling always leads to a better the energy transfer efficiency. The effect of back energy transfer (lanthanide to organic ligands) was also investigated in Tb clusters. It was revealed that there is an activation energy in the energy transfer between lanthanide and organic ligands, and the frequency factor that determine the coupling strength between the two excited states.

Finally, the thesis combines the previous three conclusions and investigates the effect of energy transfer between $Tb^{3+}$ ions on back energy transfer in nonanuclear Tb clusters. The striking result was that the energy transfer between $Tb^{3+}$ ions mitigates the effect of back energy transfer by transferring the energy to a more efficiently emitting Tb site. It was also found that the effect is especially prominent for clusters due to its structure. The thesis provides a new insight that lanthanide clusters are potentially a new class of efficiently luminescent functional material, as well as fundamental insights into energy transfer processes in lanthanide complexes. The methodology is also thoroughly explained in the thesis that will serve as a useful reference for those working in this field of research.

Sapporo, Japan
August 2018

Prof. Dr. Yasuchika Hasegawa

**Parts of this thesis have been published in the following journal articles:**

1. Omagari S, Nakanishi T, Kitagawa Y, Seki T, Fushimi K, Ito H, Meijerink A, Hasegawa Y (2018) Spin-orbit Coupling Dependent Energy Transfer in Luminescent Nonanuclear Yb-Gd / Yb-Lu Clusters. J Lumin 201:170–175. https://doi.org/10.1016/j.jlumin.2018.04.049
2. Omagari S, Nakanishi T, Hirai Y, Kitagawa Y, Seki T, Ito H, Hasegawa Y (2018) Origin of Concentration Quenching in Ytterbium Coordination Polymers: Phonon-Assisted Energy Transfer. Eur J Inorg Chem 561–567. https://doi.org/10.1002/ejic.201701396
3. Omagari S, Nakanishi T, Hasegawa Y et al (2016) Critical Role of Energy Transfer Between Terbium Ions for Suppression of Back Energy Transfer in Nonanuclear Terbium Clusters. Sci Rep 6:37008. https://doi.org/10.1038/srep37008
4. Omagari S, Nakanishi T, Kitagawa Y, Seki T, Fushimi K, Ito H, Meijerink A, Hasegawa Y (2015) Effective Photosensitized Energy Transfer of Nonanuclear Terbium Clusters Using Methyl Salicylate Derivatives. J Phys Chem A 119:1943–1947. https://doi.org/10.1021/jp512892f
5. Omagari S, Nakanishi T, Kitagawa Y, Fushimi K, Hasegawa Y (2015) Synthesis and Photoluminescence Properties of Nonanuclear Tb(III) Clusters with Long Alkyl Chain Group. e-J Surf Sci Nanotech 13:27–30. https://doi.org/10.1380/ejssnt.2015.27

# Other publications by the author:

1. Dalafu H A, Rosa N, James D, Asuigui D R C, McNamara M, Kawashima A, Omagari S, Nakanishi T, Hasegawa Y, Stoll S L (2018) Solid-State and Nanoparticle Synthesis of $EuS_xSe_{1-x}$ Solid Solutions. Chem Mater 30:2954–2964. https://doi.org/10.1021/acs.chemmater.8b00393
2. Morisue M, Omagari S, Ueno I, Nakanishi T, Hasegawa Y, Yamamoto S, Matsui J, Sasaki S, Hikima T, Sakurai S (2018) Fully Conjugated Porphyrin Glass: Gollective Light-Harvesting Antenna for Near-Infrared Fluorescence beyond 1 μm. ACS Omega 3:4466–4474. https://doi.org/10.1021/acsomega.8b00566
3. Fiaczyk K, Omagari S, Meijerink A, Zych E (2018) Temperature Dependence of $4f^{n-1}5d^1 \rightarrow 4f^n$ Luminescence of $Ce^{3+}$ and $Pr^{3+}$ Ions in $Sr_2GeO_4$ Host. J Lumin 198:163–170. https://doi.org/10.1016/j.jlumin.2018.02.034
4. Seki T, Tokodai N, Omagari S, Nakanishi T, Hasegawa Y, Iwasa T, Taketsugu T, Ito H (2017) A Luminescent Mechanochromic 9-Anthryl Gold(I) Isocyanide Complex with an Emission Maximum at 900 nm after Mechanical Stimulation. J Am Chem Soc 139:6514–6517. https://doi.org/10.1021/jacs.7b00587

5. Watanabe A, Kobayashi A, Satoh E, Nagao Y, Omagari S, Nakanishi T, Hasegawa Y, Sameera W M C, Yoshida M, Kato M (2017) Development of Ion-Conductive and Vapoluminescent Porous Coordination Polymers Composed of Ruthenium(II) Metalloligand. Inorg Chem 56:3005–3013. https://doi.org/10.1021/acs.inorgchem.6b03123
6. Murakami K, Ooyama Y, Watase S, Matsukawa K, Omagari S, Nakanishi T, Hasegawa Y, Inumaru K, Ohshita J (2016) Chem Lett 45:502–504. https://doi.org/10.1246/cl.160036
7. Kataoka H, Nakanishi T, Omagari S, Takabatake Y, Hasegawa Y (2016) Drastically Improved Durability and Efficiency of Silicon Solar Cells Using Hyper-Stable Lanthanide Coordination Polymer Beads. Bull Chem Soc Jpn 89:103–109. https://doi.org/10.1246/bcsj.20150325
8. Kataoka H, Omagari S, Nakanishi T, Hasegawa Y (2015) Photo-degradation Analysis of Luminescent Polymers with Lanthanide Complexes. J Photopolym Sci Technol 28:247–254. https://doi.org/10.2494/photopolymer.28.247
9. Kataoka H, Omagari S, Nakanishi T, Hasegawa Y (2015) EVA Thin Film with Thermoo- and Moisture-stable Luminescent Copolymer Beads Composed of Eu(III) Complexes for Improvement of Energy Conversion Efficiency on Silicon Solar Cell. Opt Mater 42:411–416. https://doi.org/10.1016/j.optmat.2015.01.038

# Acknowledgements

Throughout my five years that I have been in the Advanced Materials Chemistry Laboratory (AMC Laboratory), there were many great experiences that changed my life. Such experiences are not limited to research, but social life, friendship, responsibility, hardships, etc., as a matter of fact, everything. The people of the AMC Laboratory become the central part of my life and have influenced me in every way possible. It is thus inevitable that my Ph.D. degree had tremendous contributions from the people around me. Below is my attempt at showing my greatest appreciation toward those that have contributed to my Ph.D. thesis.

First and foremost, I would like to express my sincere gratitude to my main advisors, Prof. Yasuchika Hasegawa, Ph.D., and Prof. Takayuki Nakanishi, Ph.D., of the AMC Laboratory. Professor Nakanishi in particular, my primary advisor, has supervised me for the five years that I was in the laboratory (including undergraduate). I am proud to be the first one to graduate the master's course under him, and now the Ph.D. course. In terms of research, he and I had a very different approach, and yet, he still granted me the freedom to choose the style I feel I am good at. Professor Hasegawa is the reason why I decided to join this laboratory in the first place. I thank him for granting me the place to do what I can do. My advisors were very supportive of my academic career and goals. They were also the ones that stopped me when I have gone too far with my curiosity, which was crucially important to produce meaningful results. This thesis would never have even begun without them.

I would also like to express my gratitude toward my other advisors, Prof. Koji Fushimi, Ph.D., and Prof. Yuichi Kitagawa, Ph.D., of the AMC Laboratory as well as Prof. Noboru Kitamura, Ph.D., of the Analytical Chemistry Laboratory. I thank the three for their academic discussions. It has greatly helped me approach science. Professor Fushimi's guidance in academic writing was also crucial for polishing my works. Quantum chemical discussions of excited states with Prof. Kitagawa and Prof. Kitamura gave me a good sense of scientific understanding of various photophysical phenomena.

I wish to thoroughly thank Prof. Andries Meijerink, Ph.D., of Condensed Matter and Interfaces (CMI) group at Utrecht University (the Netherlands) for letting me stay in his group for two months. He was always with a smile and did not hesitate whatsoever to help me even if he was extremely busy.

I would also like to express my gratitude toward Prof. Hajime Ito, Ph.D., and Prof. Tomohiro Seki, Ph.D., of Organoelement Chemistry Laboratory. They were very supportive of my research and even provided me the opportunity to collaborate.

Next, I also greatly thank Hiroyasu Sato, Ph.D., of Rigaku Corporation. He was quick to help when I need them the most in the crystallographic analyses.

I also deeply appreciate the support of the Japan Society for the Promotion of Science (JSPS) with their Research Fellowship for Young Scientists (DC1) program. The program helped me financially and allowed me to focus on my research.

I wish to thank my fellow Ph.D. course colleagues of the same grade, Satoshi Wada, Masanori Yamamoto, and Yudai Yamamoto, as well as my seniors: Yuichi Hirai, Ph.D., Akira Kawashima, Ph.D., Jun Seob Lee, Ph.D., Yu Takabatake, Ph.D., Kei Yanagisawa, Ph.D. The way we supported each other was one of the main reasons for my intellectual improvements. I remember when I first came to this laboratory, completely clueless about what we were doing and attempting to understand all these new knowledge. Fortunately, these people were there to selflessly help me. Hopefully, we will stay in touch even after we part ways. I wish you guys the best. I would like to continue on and thank all the other members of the AMC Laboratory. I must also mention those in the CMI Laboratory during my stay in the Netherlands.

I thank Hokkaido University, the place where it all began. I spent eight years in this university, and I cannot imagine what would have happened if I went anywhere else. Eight years is actually the longest that I have ever lived in one place, and thus, I can say that Sapporo has now effectively became my hometown. This place has provided me with the most precious experience in all aspects and made me who I am.

Last but not least, I cannot forget to thank my mom and dad. I know I am not the most obedient child you have, and we tend to have a different opinion on things (even today and probably forever). Nonetheless, your support was without a doubt crucial to get to this point in my life. I also thank my brother and sister who were with great talents. Your existence and your success motivate me in doing what I do.

There were good times and bad times. The bad times strengthened me to withstand all hardships, whether I deal with it alone or with the support of others. The good times provided me a reason to live. I thank this world for the beautiful and the ugly things in my life.

# Contents

# Abbreviations

| | |
|---|---|
| BET | Back energy transfer (Ln→$T_1$ ET) |
| ET | Energy transfer |
| FET | Forward energy transfer ($T_1$→Ln ET) |
| ISC | Intersystem crossing |
| KS | Killer Site |
| Ln | Lanthanide |
| LnLnET | Ln-to-Ln energy transfer |
| NIR | Near infrared |
| PAETR | Phonon-assisted energy transfer relaxation |
| $S_0$ | Singlet ground state |
| $S_1$ | Singlet excited state |
| SOC | Spin–orbit coupling |
| $T_1$ | Triplet excited state |
| UV | Ultraviolet |

# Chapter 1
# General Introduction

**Abstract** Photoluminescent materials are used everywhere in our current society from daily lives to industries and research. As simple as it may seem, there are numerous types of photoluminescent materials (each with possibly different luminescence mechanisms), and tremendous amount of effort has been invested in research on these materials, sometimes going far into the fundamentals of quantum mechanics. The thesis focuses on lanthanide complexes, which is only one of the many types of luminescent materials available today. Understanding the contents of this thesis requires that the reader understands the underlying physical and chemical concepts of photoluminescence, lanthanides, and lanthanide complexes. This chapter provides the most basic concepts that the readers are required to understand. First, a very brief history of luminescent materials and their applications are mentioned. Then, the qualitative descriptions of photons (light) and the quantum mechanical descriptions of photoluminescence are introduced with minimal amount of mathematics. The chemical and physical properties of lanthanides and lanthanide complexes are explained along with several examples of lanthanide complexes. Finally, the outline of this thesis is provided.

## 1.1 Introduction

"Light brings us the news of the Universe," said W. H. Bragg in The Universe of Light in 1933 [1]. He was indeed right that it does. Production, manipulation, and utilization of light have been one of the most extensively studied areas in science, and also one of the most practical. Humanity has become increasingly reliant on technology utilizing light throughout the history, and it is now difficult to find a society that does not rely on light. Plants (and even some bacteria) also absorb sunlight for photosynthesis, which is vital to the ecosystem including our existence. Utilization of light has also lead to other scientific discoveries, with one of the most recent (2017) Nobel Prize in Physics going to R. Weiss, K. Throne, and B. Barish for the observation of gravitational waves using laser interferometer gravitational-wave observatory (LIGO).

 
S. Omagari, *Energy Transfer Processes in Polynuclear Lanthanide Complexes*, Springer Theses, https://doi.org/10.1007/978-981-13-6049-7_1

The history of science and technology based on light is quite long and in a sense could date all the way back to the Stone Age when human had used fire for heat and light. However, modern (commercialized) lighting was only available since the early 20th century with the introduction of incandescent light bulbs. Today, fluorescent lamps and light-emitting diode (LED) lamps are usually used for general lighting, but high-pressure sodium lamps are commonly used for street lights. Cathode ray tube (CRT) based displays are rarely produced now and are replaced by liquid crystal (LCD) and (briefly) plasma displays. More recently, organic light-emitting diode (OLED) displays have emerged as another type of displays. Lasers (light amplification by stimulated emission of radiation) are also one of the most important technologies that use light, being invented by T. H. Maiman in 1960. Applications include, but not limited to, telecommunication, quantum computing, optical memory, medicine, military weapons, and cutting/welding in industry. As a tool for the further development in science, lasers are used in time-resolved spectroscopy, Fourier-transform infrared spectroscopy (FTIR), Raman spectroscopy, fluorescence microscopy, dynamic light scattering (DLS), laser fusion, laser cooling, and ion trapping.

### *1.1.1 Descriptions of Light*

There are several quantitative descriptions of light (or more precisely, electromagnetic radiation), but the two main descriptions are intensity and energy of the light. Some other descriptions are coherence and polarity, but it is not the focus of this thesis. The difference between the intensity and energy of light can be visually explained by particle-like interpretation of light called "photon" (Fig. 1.1). The energy of light refers to photon energy, meaning the energy carried by a single photon. The unit used can be wavenumber [$cm^{-1}$], electronvolt [eV], or wavelength [nm], all of which can be easily converted. As shown in the photon energy spectrum in Fig. 1.1a, light is classified into radio waves, (microwaves), infrared light, visible light, ultraviolet light, X-rays, and gamma rays depending on the energy carried by a photon. The visible region and the regions below and above the energy of this region (near-infrared and ultraviolet) is of particular importance in this thesis. As the name implies, the visible region of light is the region that the human eyes can detect. A color of light could be composed of light of single energy (color) or combination of light of different energies. The primary colors (Fig. 1.2) are the (minimum number of) pure colors to be able to express any color by a combination of them. To express a certain color, one can use additive or subtractive mixing of primary colors. In additive mixing, one combines light of different energies. For example, a total mixture of the conventional red, green, and blue (RGB) primary colors yields white (Fig. 1.2a). In subtractive mixing, one combines pigments that absorb light of different energies. For example, a total mixture of yellow, magenta, and cyan pigments yields black by absorbing all colors of light, rendering it unable for the human eyes to receive any light (Fig. 1.2b). Near-infrared (NIR) and ultraviolet (UV) lights are not visible to human eyes, but

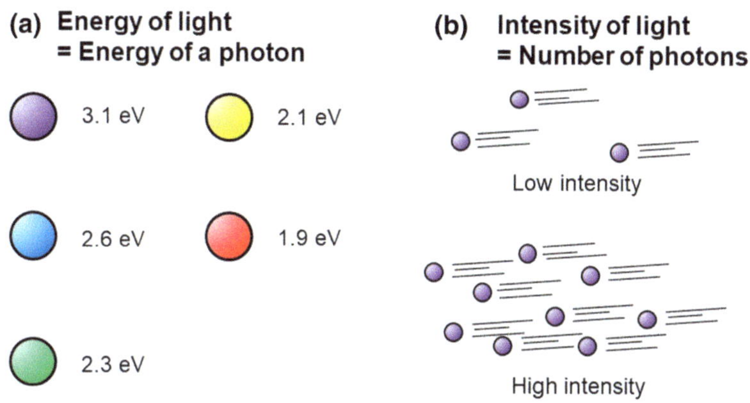


**Fig. 1.1** **a** *Energy* and **b** *intensity* of the light in the particle (rough) description

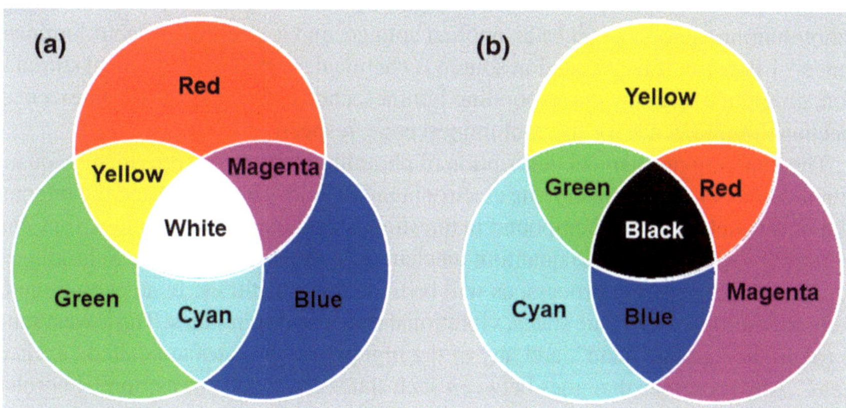


**Fig. 1.2** **a** Additive mixing (primary colors: blue, green, and red), and **b** subtractive mixing (primary colors: cyan, yellow, and magenta)

there are many applications. NIR light is used for bioimaging for their high transmission in biological tissues, and opto-telecommunication for their low loss in optical fibers. Most compounds absorb light of UV region. Therefore, they can be used for forensic analysis and drug detection. High energy UV is potentially hazardous to organisms as it can disrupt DNA molecules, and can be used as disinfectants.

The intensity of light roughly refers to the number of photons that are detected by the observer (Fig. 1.1b). However, most units used to describe the intensity of light do not count the number of photons due to the instrumental difficulty in precisely detecting the number of photons (of single energy). Most detectors measure the total energy resulting from the photons that enter the detector instead. The most common unit used to describe the intensity of light is irradiance (also known as flux density) expressed in W $m^{-2}$ (watts per meter squared), which is radiant flux received by a surface of unit area. Radiant flux, expressed in W (or J $s^{-1}$), refers to the total energy of light (photons) emitted per unit time. As can be inferred from the definition of radiant flux, irradiance includes information on the energy of light and not just the intensity of light. In general, a light may be composed of photons with various energy. In spectroscopy, where one measures the intensity of each photon energy, it is often a poor practice to judge the intensity of light based on the irradiance without the information of the photon energy that the light is composed of.

### *1.1.2 Photoluminescence*

A material that can emit photons is called luminescent material. Emission of photons requires that the material is electronically excited by some excitation source. Depending on the excitation source, a prefix is added to the word "luminescence". For example, if a material is excited by light, the resulting luminescence is termed "photoluminescence". Let it be an applied voltage, and it is termed "electroluminescence". Likewise, if the excitation source is chemical reactions, mechanical stimulation, or applied heat, the corresponding luminescence is called chemiluminescence, mechanoluminescence, or thermoluminescence, respectively.

The quantum mechanical description of photoluminescence is provided by quantum field theory. In this theory, the electromagnetic field is quantized and the interaction of this field with the compound in question is calculated. Nonetheless, they can be roughly approximated to quantum mechanics in the non-relativistic Schrödinger equation for one photon process, as will be treated in this thesis. In any compounds there are multiple electronic states, vibrational states, and spin states. The lowest state is called the "ground state", and any of the higher energy states are called "excited state". Any processes that goes between such states are called "transition". Photoluminescence in this case is the transition between two electronic states involving an absorption and emission of a photon corresponding to the energy gap between the states. Under adiabatic approximation (such as Born-Huang, Born-Oppenheimer, or crude adiabatic), the movement of electrons is assumed to move much faster than the nuclei so that vibrational states are determined once an electronic state for a given configuration of the nuclei of the compound is solved. This produces a potential energy surface, and for diatomic molecules a potential energy curve (Fig. 1.3a). Both the potential energy of ground and excited states are depicted as quadratic curves. This is because for a diatomic compound, two atoms can be considered to be bonded together like a spring, classically speaking. By the Hooke's law, this produces a quadratic potential curve (Fig. 1.3b). The $x$-axis (nuclear coordinate) position of the minimum of the potential energy curves is called the "equilibrium position". The equilibrium position of the two potential curves is depicted as being different because generally the different electronic states usually involve changes to the bonding length and strength, and thus the energetically stable point as well. Quantum mechanically, vibrations of the molecules are quantized (quantum harmonic oscillator) and thus there are equal gaps in energy between the vibrational states. In reality, a compound is comprised of much more than two atoms, which increases the degree of freedom

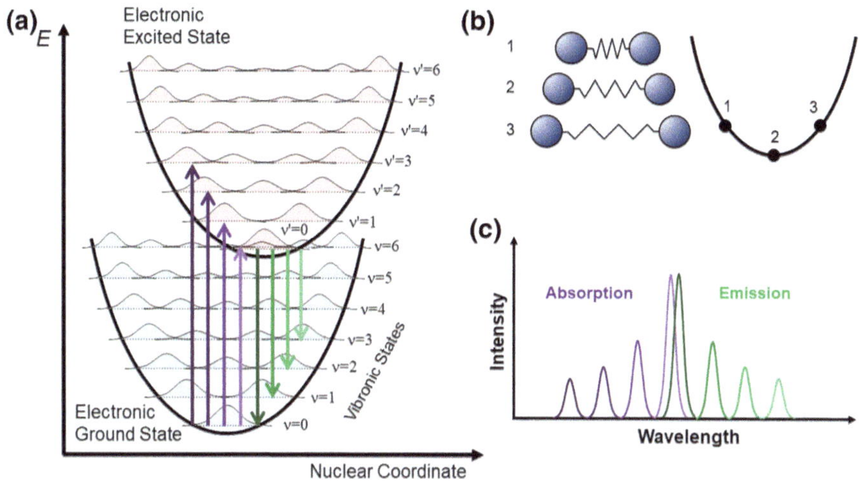


**Fig. 1.3** **a** Potential energy curve for a diatomic compound and photoluminescence. Up arrow represents absorption and down arrow represents emission. The waves on each of the potential depicts the probability density of the vibronic modes. ν represents the quantum number of the vibronic mode. **b** Classical depiction of the vibration o a two-atom molecule and their corresponding energies on the potential curve. **c** Absorption and emission spectra resulting from **a**). Wavelength is the inverse of energy. Longer wavelength means lower energy photon and shorter wavelength means larger energy photon

in the movement of the nuclei, resulting in multiple vibrational modes. Moreover, the potential energy surface is not precisely quadratic, especially in higher energy limit because bonds can significantly change (anharmonic).

Electronic transition (or luminescence) typically proceeds from the lowest vibrational states of an electronic state (higher vibrational states can still be populated by thermal excitation). Under Franck-Condon approximation (which assumes Born-Oppenheimer approximation) an electronic transition is fully vertical as shown in the arrows of Fig. 1.3a because the slow moving nuclei cannot change its configuration faster than the electrons. In practice, absorption and emission spectra, which show the dependence of photon energy on the intensity of the transition, are in the form of that shown in Fig. 1.3c. The vibrational bands are observed because the electromagnetic field couples vibrational states of an electronic state with another electronic state, and the emission intensity depends on the amplitude of the overlap of vibrational states. The degree of coupling depends on the displacement of the equilibrium position of the ground and excited state.

All photophysical processes are kinetic. This means that luminescence can be expressed in terms of how "fast" they emit a photon from an excited state. The rate constant of this process is called "radiative rate constant" and is in the unit of emission of photons per second [$s^{-1}$]. Many of the light sources seem like they are continuously emitting. Microscopically, if only one single emitting center is considered, blinking of an emitting center at a rate mentioned above is observed

[2]. Macroscopically, numerous emitting centers blink that make the material looks as if it is continuously emitting. The efficiency of photoluminescence of a material is expressed by "quantum yield" Φ defined by the number of emitted photons per number of absorbed photons:

$$\Phi = \frac{(\text{Number of emitted photons})}{(\text{Number of absorbed photons})}. \tag{1.1}$$

In practice, the quantum yield of a photoluminescent material is rarely unity (100%). This is because excited material can deactivate to a ground state without photon emission in the form of heat through vibration of the compound and/or the surrounding matrix. This phenomenon is called internal conversion (but often confusingly called "vibrational relaxation") [3].

Internal conversion is a nonadiabatic process. This means that the adiabatic approximation used in the case of luminescence cannot be directly applied. Quantum mechanically, the off-diagonal matrix elements resulting from the nuclear kinetic energy operator cause coupling of vibrational and electronic states. The nonradiative rate constant of this process is exponentially dependent on the energy gap between two electronic states. This is the "energy gap law", and it is also the basis of Kasha's rule [4]. Kasha's rule states that an emission will occur from the lowest vibrational state of the lowest excited electronic state (with some exceptions like azulenes). In many molecules, the energy gap between two electronically excited states is too small that internal conversion between these two states proceed much faster than luminescence from the second excited state. In contrast, energy gap between the first excited and ground states are usually larger allowing luminescence. The implication of this rule is that NIR luminescence (emission of a photon of low energy) is typically more inefficient than UV-visible luminescence (emission of a photon of high energy).

Spin multiplicity is another factor that greatly affects the emission process. A transition between two electronic states of the same spin multiplicity is an allowed process and is called "fluorescence." On the other hand, a transition between those of different multiplicity is a forbidden process and is called "phosphorescence." The term "forbidden" refers to the fact that to the zeroth-order, the transition matrix element is zero, meaning that the transition cannot happen. In practice, they are still observed due to mixing with other states of allowed transition and/or by nonadiabatic coupling. Typically, fluorescence have a large radiative rate constant ranging from $10^7$ to $10^{11}$ $s^{-1}$ while phosphorescence has a small radiative rate constant ranging from $10^{1-1}$ to $0^4$ $s^{-1}$. Consequently, phosphorescence is usually not observed in room temperature as internal conversion dominates.

There are numerous types of photoluminescent materials available today, but the common (and commercially available) ones used today include Ce: $YPO_4$ (blue), Tb: $YPO_4$ (green), Eu: $Y_2O_3$ (red) (these three are used for fluorescent lamps), and Ce: $Y_3A_{15}O_{12}$ (green-yellow, used for white LED). These materials can be considered as spectral converters, which convert light of certain wavelength (or energy) into the light of different wavelength. Other potential applications of spectral converters are films to produce an optimal wavelength of light for solar panels and crops. Some

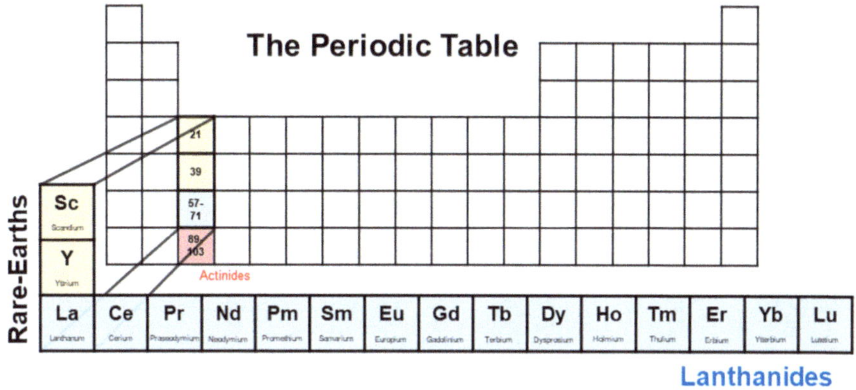


**Fig. 1.4** Rare-earths (shaded in yellow and blue), lanthanides (blue), and actinides (red) in the periodic table. The numbers are the atomic numbers

photoluminescent materials are used as solid-state laser medium such as Ti-sapphire lasers, ruby lasers, and Nd: YAG lasers.

## 1.2 Lanthanides

Lanthanides (Ln) are 15 metallic elements in the periodic table with atom number 57–71 (Fig. 1.4, blue elements). They are also called "rare-earth" along with scandium (Sc) and yttrium (Y) for their similar chemical properties (Fig. 1.4, yellow and blue elements). The name rare-earth originated from their low availability at the time they were discovered (1787) due to the difficulty in purification. Lanthanides could also be classified as "f-elements" along with actinides because electron gradually fills seven 4f-orbitals (5f-orbitals for actinides) with increasing atomic number (Fig. 1.4, blue and red elements). Contrary to its classification name "rare earth", they are relatively abundant in Earth's crust. Cerium (Ce) is the 35th most abundant element in Earth's crust. The least abundant (or nonexistent) is promethium (Pm) in which there is no stable isotope available and is produced artificially by nuclear reaction. Rare earth as a whole is estimated to be up to 242 ppm in the Earth's crust, more than the amount of carbon (200 ppm) [5].

The filling of seven 4f-orbitals (Fig. 1.5a) [6] in the lanthanides gives rise to many unique physical properties. The chemical properties of lanthanides have also been considered important in synthetic chemistry as a catalyst. General chemical properties of lanthanides include similar ionic radii, reduction potentials, oxidization states, and electronegativity. Physical properties of lanthanides include luminescence and magnetism, which both are intrinsic and usually remain rather unaffected by outer environment. Lanthanides have been playing an important role in the current technologies such as lasers, radars, telecommunication, lighting equipment, and MRIs

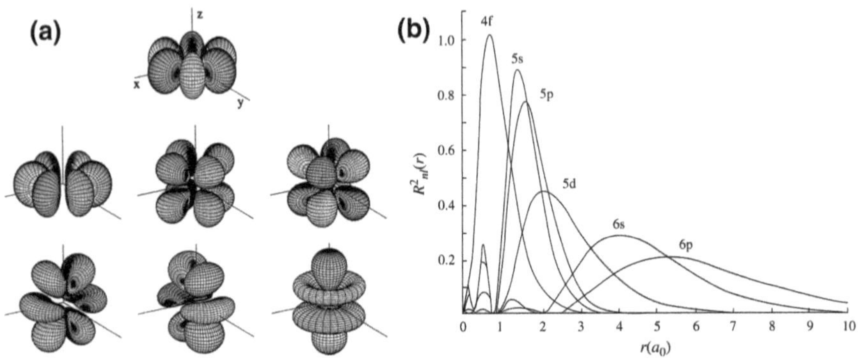


**Fig. 1.5** **a** Depiction of the real wavefunctions of 4f-orbitals reprinted from Ref. [6] with permission from Springer Publishing Company. **b** Radial part of the wave function of the orbitals reprinted from Ref. [5] with permission from Wiley

[7]. This section will mainly describe the basic properties of lanthanide ions (Ln(III) ions) with greater emphasis on the photophysical properties.

### 1.2.1 Electron Configuration and Chemistry

The electron configuration of lanthanide is the main reason for their chemical and physical properties. The electron configuration of lanthanides (as well as Sc and Y) is shown in Table 1.1. In its metallic state (0 valence state), the 6s-orbital is fully occupied, and sometimes a single electron occupies the 5d-orbital as well. However, lanthanides in its metallic state are quite reactive (their reactivity is similar to alkali earth metal) and quickly oxidize to a trivalent state. In the trivalent state, lanthanide (Ln(III)) loses three electrons: two from the 6s-orbital and one from either the 4f-orbital or 5d-orbital if it was occupied. This results in a Ln(III) ion with 4f-orbitals that is not completely filled (except lanthanum (La), which do not have any 4f electrons, and lutetium (Lu), which is fully filled). As shown in Fig. 1.5b, 4f-orbitals are narrowly distributed closer to the nucleus than the 5s, 5p, 5d, and 6s-orbitals [5]. Therefore, 4f-orbitals are "protected" by the outer orbitals and do not contribute to bonding. As a result, the chemical and physical properties are similar throughout the lanthanide series.

Lanthanides are commonly in their trivalent state, but under some conditions, divalent and quadrivalent states are also possible. La and Lu have none and fully occupied 4f-orbitals, respectively, and thus are stable electron configurations. Gadolinium (Gd) has seven 4f-electrons making all of the 4f-orbitals half filled, and is in a relatively stable electron configuration as well. Ce and terbium (Tb) take the relatively stable quadrivalent state by taking the same electron configuration as La and Gd, respectively. Similarly, europium (Eu) and ytterbium (Yb) also reduce into a divalent state

**Table 1.1** Electron configuration, ionic radii, and electronegativity of lanthanides

| Element | Atomic Number | Configuration of atom | Configuration of trivalent ion | Ionic radii/Å[a] | Electro-negativity[b] |
|---|---|---|---|---|---|
| Scandium Sc | 21 | $[Ar]3d^1 4s^2$ | [Ar] | 0.87 | 1.20 |
| Yttrium Y | 39 | $[Kr]4d^1 5s^2$ | [Kr] | 1.019 | 1.11 |
| Lanthanum La | 57 | $[Xe]5d^1 6s^2$ | [Xe] | 1.16 | 1.08 |
| Cerium Ce | 58 | $[Xe]4f^1 5d^1 6s^2$ | $[Xe]4f^1$ | 1.143 | 1.08 |
| Praseodymium Pr | 59 | $[Xe]4f^3 6s^2$ | $[Xe]4f^2$ | 1.126 | 1.07 |
| Neodymium Nd | 60 | $[Xe]4f^4 6s^2$ | $[Xe]4f^3$ | 1.109 | 1.07 |
| Promethium Pm | 61 | $[Xe]4f^5 6s^2$ | $[Xe]4f^4$ | 1.093 | 1.07 |
| Samarium Sm | 62 | $[Xe]4f^6 6s^2$ | $[Xe]4f^5$ | 1.079 | 1.07 |
| Europium Eu | 63 | $[Xe]4f^7 6s^2$ | $[Xe]4f^6$ | 1.066 | 1.01 |
| Gadolinium Gd | 64 | $[Xe]4f^7 5d^1 6s^2$ | $[Xe]4f^7$ | 1.053 | 1.11 |
| Terbium Tb | 65 | $[Xe]4f^9 6s^2$ | $[Xe]4f^8$ | 1.04 | 1.10 |
| Dysprosium Dy | 66 | $[Xe]4f^{10} 6s^2$ | $[Xe]4f^9$ | 1.027 | 1.10 |
| Holmium Ho | 67 | $[Xe]4f^{11} 6s^2$ | $[Xe]4f^{10}$ | 1.015 | 1.10 |
| Erbium Er | 68 | $[Xe]4f^{12} 6s^2$ | $[Xe]4f^{11}$ | 1.004 | 1.11 |
| Thulium Tm | 69 | $[Xe]4f^{13} 6s^2$ | $[Xe]4f^{12}$ | 0.994 | 1.11 |
| Ytterbium Yb | 70 | $[Xe]4f^{14} 6s^2$ | $[Xe]4f^{13}$ | 0.985 | 1.06 |
| Lutetium Lu | 71 | $[Xe]4f^{14} 5d^1 6s^2$ | $[Xe]4f^{14}$ | 0.977 | 1.14 |

a) Shannon Radii of 8-coordinate Ln(III) ion
b) Value of trivalent ion in Pauling scale

by possessing the electron configuration of Gd and Lu, respectively. These divalent and quadrivalent states are not stable and require synthesis in a controlled condition.

Relatively large ionic radii of Ln(III) ions compared to other metal ions give rise to variable coordination number of ligating atoms ranging from 3 to 12. Ionic radii of Ln(III) ions gradually contract from 1.16 Å of La to 0.977 Å of Lu, known as the "lanthanide contraction." This phenomenon occurs due to the increase in the number of electrons in the inner orbitals that cannot adequately shield the increasing positive charge of nuclei, causing the outer orbitals to be drawn closer to the nuclei. The slight difference in ionic radii is important in purification and extraction of each Ln(III) ions from oxides. However, the effect of lanthanide contraction is small enough to maintain their similarity in chemical properties and thus considered to have similar ionic radii [6, 8].

### 1.2.2 Luminescence

Ln(III) ions show distinctive luminescence properties mainly deriving from the 4f-orbitals. This includes spectrally sharp absorption and emission bands (similar to atomic spectra), long emission lifetimes, and small absorption coefficients. This subsection covers basic luminescence properties of Ln(III) ions, namely the energy levels, the spectral width of the emission bands, and emission lifetimes. Further details of what is explained below are provided in Chap. 2.

Energy of 4f-states in Ln(III) ions in its free ion state is fixed for each Ln(III) ions, and thus the absorption and emission wavelength of Ln(III) ions are fixed [9–12]. The electronic states are determined by electrostatic coupling and spin-orbit coupling of 4f-electrons to the first-order. As stated before, 4f-orbitals are shielded by 5s-, 5p-, 5d-, and 6s-orbitals, and do not contribute to bonding. Therefore, unlike the d-orbitals in transition metals, the crystal-field is a minor perturbation in the determination of the 4f-states. The energy level of 4f-states is shown in the "Dieke diagram" in Fig. 1.6a [13]. Each energy level can be labeled by a Russell-Saunders term symbol: $^{2S+1}L_J$, where $S$, $L$, and $J$ are the total spin quantum number ($2S + 1$ represents spin multiplicity), total orbital quantum number, and total angular quantum number, respectively. Combination of $S$, $L$, and $J$ gives a level, and each level contains $2J + 1$ possible microstates. These intrinsic energy levels are advantageous for designing a luminescent material as one can easily expect what wavelength the material is excited and emits.

The absorption and emission of Ln(III) ions are spectrally sharp, similar to an atomic spectrum [14, 15]. Figure 1.6b shows the potential energy of two 4f-states in comparison with a typical luminescent center. Since the 4f-orbitals do not contribute to bonding, the shape of the potential curve, which is determined by bonding and position of the atoms, is unaffected for all 4f-states. Furthermore, the lack of involvement in the bonding keeps the vibrational coupling (a.k.a. electron-phonon coupling) of 4f-states very weak [5, 16]. As a result, absorption and emission between two 4f-states are sharp. The splitting usually observed in a single band of Ln(III) ions is called Stark splitting, caused by a crystal-field that lifts some of the degeneracy of $2J + 1$ microstates. Thus, a different spectral shape is observed for different coordination environment. Sometimes, the vibronic structure is observed in Ln(III) absorption and emission, called phonon-sidebands [17–19]. Nonetheless, these bands are very low in intensity.

The emission lifetimes of Ln(III) ions are typically long; they are in the order of microseconds to milliseconds. Emission lifetime $\tau_{obs}$ in its simplest form is given by the following equation:

$$\tau_{obs} = \frac{1}{kr + knr}, \tag{1.2}$$

where $kr$ and $knr$ are radiative and nonradiative rate constants, respectively. In lanthanides, both of these rate constants are small in the order of $10^2$–$10^3$ $s^{-1}$. The $kr$ of Ln(III) ions is small, meaning the electronic transitions are weak (small oscillator

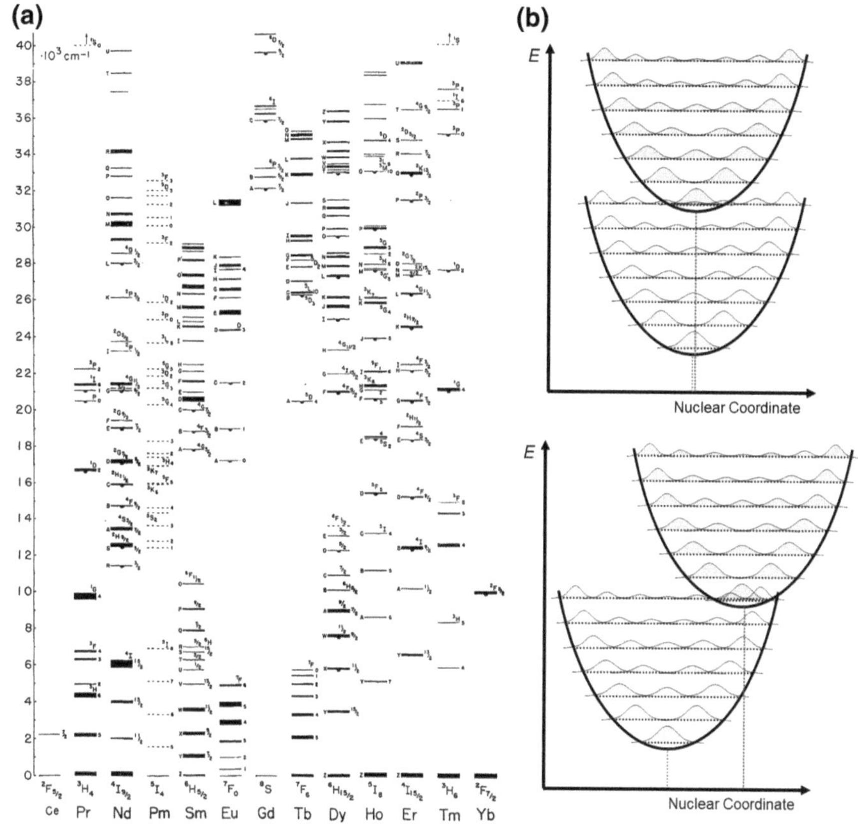


**Fig. 1.6** **a** Dieke diagram representing 4f-state energy levels. Reprinted with permission from Ref. [13], The Optical Society. **b** Schematic image of the potential energy curve of the ground and excited 4f-states (top) and typical luminescent center (bottom)

strength), due to the 4f-4f transitions being Laporte forbidden [6, 20]. Absorption and emission can still be observed in practice because crystal-field causes a mixture of 4f5d-states into 4f-states, leading to partially allowed 4f-4f transitions. Absorption coefficient is no more than 10 $M^{-1}$ $cm^{-1}$ and in most cases smaller than 1 $M^{-1}$ $cm^{-1}$. Qualitatively, the oscillator strength can be made stronger by distorting the symmetry of ligand field (coordination geometry) [21, 22]. The oscillator strength of dipole transitions in a Ln(III) ion can be quantitatively calculated using semi-empirical Judd-Ofelt theory [23, 24]. The *knr* (relaxation via vibrational states) is also small in 4f-4f transitions. As mentioned in the paragraph above, the shape and the nuclear configurational position of the potential curves between 4f-states are nearly identical and the electron-phonon coupling is thus very weak. This means that the relaxation from the crossing point between potentials is low in probability and relaxation through overlap with the vibrational modes is very small as well. Additionally, the wave function of the vibrational mode of the ground and excited states tends to be smaller for higher vibrational states of the ground state, further reducing the overlap.

The small absorption coefficient of Ln(III) ions makes it impractical to use the 4f-4f transition for exciting the ions directly. Only few Ln(III) ions such as Tb(III) and Eu(III) ions have, respectively, 4f-4f5d and charge transfer transition in the visible to UV-violet region with relatively large absorption coefficient that can be used to excite 4f-states [6, 25]. Ln(III) ions have many intrinsic properties that distinguish themselves from other elements, making them a necessity for current and future technology. If the absorption band can be fully tuned with very large absorption coefficient, Ln(III) ions can be used efficiently and would be necessary for long-term technological advances.

## 1.3 Lanthanide Complexes

Lanthanides, as with other metal elements, form chemical bonds with nonmetal elements. Since the 4f-orbitals do not contribute to chemical bonding and the ionic radii is large, the chemical bonds with lanthanides are mainly ionic. Such property leads to Ln(III) complexes (Fig. 1.7a) forming a variety of coordination number and structure and is heavily dependent on the steric effect. Coordination number of Ln(III) complexes is primarily eight (and less commonly nine), but complexes with other coordination numbers (3–12) are reported although rare. Due to lanthanide contraction, coordination number is sometimes smaller in the heavier lanthanides even when using the same ligand. Some of the common ligands used are shown in Fig. 1.7b, namely, $\beta$-diketonate, bipyridine, DOTA (1,4,7,10-tetraaza-cyclododecane-1,4,7,10-tetraacetic acid), and PyBOX (pyridine-2,6-bisoxazoline) [5, 6, 26–32].

From the perspective of photoluminescence, one of the most important features of Ln(III) complexes is the photosensitization by organic ligands. The absorption coefficient of Ln(III) complexes is usually in the order of $10^3$–$10^4$ $M^{-1}$ $cm^{-1}$ due to the organic ligands [27]. Energy transfer (ET) from the organic ligands to Ln(III) ions efficiently produces Ln(III) excited states, leading to strong luminescence of Ln(III) ions. Additionally, functionalization is possible by the design of the ligands. For example, target-specific organic ligands can be used as Ln(III) complex for bioimaging agents [28, 33–36]. Other potential applications include lasers, lighting, OLEDs, sensors, and spectral converters [35, 37–44].

Despite the advantages of Ln(III) complexes, the quantum yield of Ln(III) complexes tends to be lower than the ceramic counterpart. This is due to the additional processes before the Ln(III) ions are excited as well as the use of organic ligands that induce vibrational relaxation [45]. In this section, the mechanism of photoluminescence of Ln(III) complexes is described first, and then the strategies to improve the quantum yield.

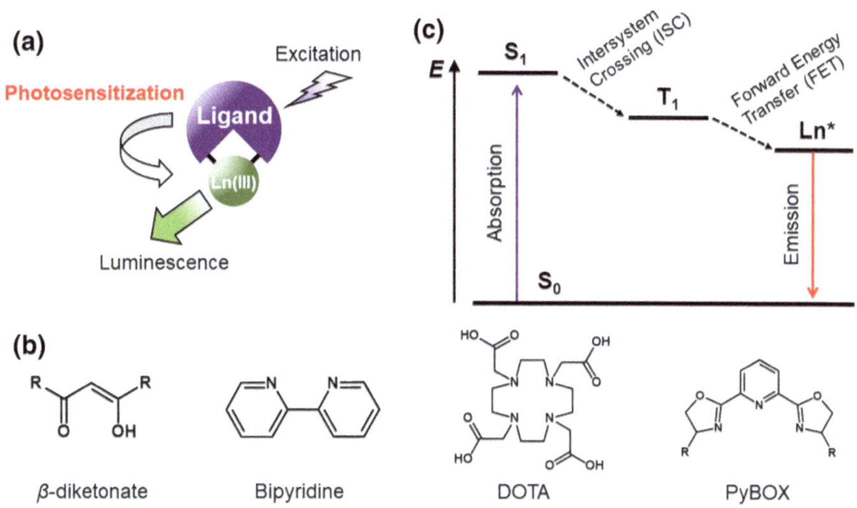


**Fig. 1.7** **a** Schematic depiction of Ln(III) complexes. **b** Common ligands used for Ln(III) complexes. **c** Luminescence mechanism of Ln(III) complexes. Ln* means an excited Ln(III) ion

### *1.3.1 Luminescence Mechanism*

Mechanism of luminescence in a typical Ln(III) complex is shown in Fig. 1.7c. First, the π-conjugated organic ligands absorb a photon to be electronically excited from the singlet ground state ($S_0$) to the singlet excited state ($S_1$). Strong spin-orbit coupling of Ln(III) ion promotes intersystem crossing (ISC) to the triplet excited state ($T_1$). This process is often said to be caused primarily by the "heavy atom effect". ET usually occurs from the $T_1$ state to an excited level of Ln(III) ion. For the convenience of the discussion throughout this thesis, $T_1 \rightarrow$ Ln ET will be denoted as "FET" (forward energy transfer), although this is not the usual convention in this area of study. With this mechanism, quantum yield of Ln(III) complexes by ligand excitation $\Phi_{\text{ligand}}$ can be calculated from the following equation:

$$\Phi_{\text{ligand}} = \eta_{\text{sens}} \times \Phi_{\text{Ln}}, \tag{1.3}$$

where $\eta_{\text{sens}}$ is the overall sensitization efficiency of the Ln(III) ion by the organic ligands, and $\Phi_{\text{Ln}}$ is the quantum yield of Ln(III) complexes by direct excitation of Ln(III) ion. $\Phi_{\text{Ln}}$ is given by the following equation:

$$\Phi_{\text{Ln}} = \frac{kr}{kr + knr}. \tag{1.4}$$

Finally, using the equation of emission lifetime $\tau_{obs}$ (Eq. 1.2) and "intrinsic" emission lifetime $\tau_{rad}$ (= $1/kr$, emission lifetime of Ln(III) in the ideal state where there are no quenching), Eq. (1.3) can be rewritten as follows:

$$\Phi_{ligand} = \eta_{sens} \times \frac{\tau_{obs}}{\tau_{rad}}. \quad (1.5)$$

As can be interpreted from Eqs. (1.3) and (1.4), improving the sensitization efficiency and raising the radiative rate constant, while reducing the nonradiative rate constant would increase the quantum yield. The strategy to enhance radiative rate constant was already explained in paragraph four of 1.2.2. Below explains the reported method to suppress nonradiative process and improve sensitization efficiency.

Nonradiative relaxation to the ground state of Ln(III) ions occurs through vibrational states. Suppressing nonradiative relaxation requires the elimination of the contributing vibrational mode(s). In the case of organic ligands, high energy vibrational modes like C–H (stretching 3000–3300 $cm^{-1}$) and O–H (stretching 3200–3700 $cm^{-1}$) can lead to a high nonradiative rate constant [21, 46]. This is especially prominent for Ln(III) ions with small energy gaps between 4f-states. To reduce nonradiative rate constant, high energy vibrational modes like C–H and O–H needs to be eliminated as much as possible. An example of the reduction by elimination of C–H vibronic mode were provided by Hasegawa et al. for Nd(III) complex [47, 48] and by Doffek et al. for Nd(III), Er(III), and Yb(III) complexes [49]. They showed that replacing C–H bonds with C–D (deuterated) bonds leads to smaller nonradiative rate constants. The suppression of nonradiative relaxation has been extensively studied.

The improvement of sensitization efficiency is complicated compared to the reduction of the nonradiative rate constant because of the multiple factors that affect the efficiency. Breaking down the sensitization efficiency, and also considering the radiative and nonradiative rate constant for $S_1$ and $T_1$ states, leads to:

$$\eta_{sens} = \eta_{ISC} \times \eta_{FET} = \frac{k_{ISC}}{kr_{S_1} + knr_{S_1} + k_{ISC}} \times \frac{k_{FET}}{kr_{T_1} + knr_{T_1} + k_{FET}}, \quad (1.6)$$

where $k_{ISC}$, $k_{FET}$, are $S_1 \rightarrow T_1$ ISC, FET rate constants, respectively. Once again, $kr$ and $knr$ are radiative and nonradiative rate constants and the subscript denotes the states in which they apply to. All of these constants influence the sensitization efficiency. Further complication here is that these rate constants are not fully independent variables, although some of them can be approximated as so. For example, spin-orbit coupling mixes states of different multiplicity, and causes $k_{ISC}$, $kr_{T_1}$, and $knr_{T_1}$ to increase while $kr_{S_1}$ and $knr_{S_1}$ to decrease.

In order to promote fast FET, an optimum level of the $T_1$ state energy in relative to the excited state of the Ln(III) ion is required. From a survey of the available reports on $T_1$ energy, the $T_1$ energy of 2000 $cm^{-1}$ above the excited state of a Ln(III) ion is optimal [50–54]. Below this, the reverse process of FET called "back energy transfer" (BET) occurs and reduces the sensitization efficiency. BET is known to

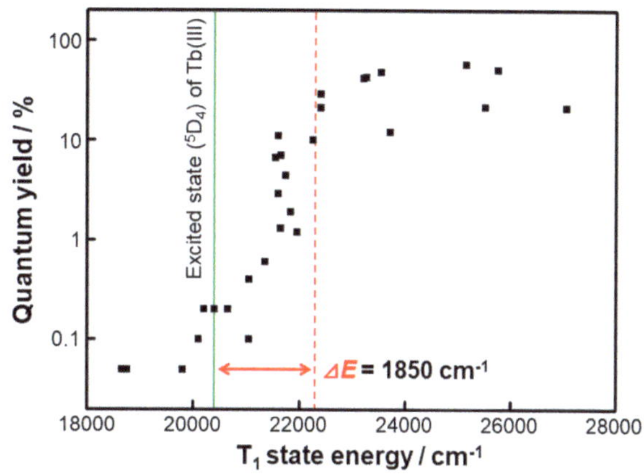


**Fig. 1.8** Relationship between the $T_1$ state energy and the quantum yield of Tb(III) complexes. Redrawn from Ref. [55] with permission from Elsevier

be particularly prominent in Tb(III) complexes, where $T_1$ state energy lower than 1850 cm$^{-1}$ significantly reduces the quantum yield (Fig. 1.8) [55]. The factors that affect the BET rate constant other than the $T_1$ state energy are not yet reported. Studying the details of the BET mechanism may provide new insight into suppressing BET and achieving high sensitization efficiency.

Raising the $S_1 \rightarrow T_1$ ISC efficiency has been largely neglected since they are considered to proceed efficiently due to the heavy atom effect of lanthanides [56, 57]. In a more general sense, $S_1 \rightarrow T_1$ ISC is induced by spin-orbit coupling, which includes the heavy atom effect and the "paramagnetic effect" (caused by the intrinsic magnetic moment produced by the 4f-electrons) [58, 59]. Reports are limited in numbers, but available reports showed that within the lanthanide series, the effect of heavy atom effect is not as pronounced as the paramagnetic effect. Guldi et al. demonstrated the effect of spin-orbit coupling on various processes within the ligands. In particular, Gd(III) complex showed a larger $S_1 \rightarrow T_1$ ISC rate constant of $3.8 \times 10^9$ s$^{-1}$ compared to that of Lu(III) complex of $1.9 \times 10^9$ s$^{-1}$, despite the latter being the heavier atom. Furthermore, the $T_1$ nonradiative process is also enhanced for the Gd(III) complex [59]. This is expected because spin-orbit coupling mixes states of different multiplicity, i.e., $S_1$ and $T_1$ states as well as $T_1$ and $S_0$ states [60].

## 1.3.2 Polynuclear Lanthanide Complexes

Polynuclear Ln(III) complex is a general term for Ln(III) complexes with multiple Ln(III) ions within a single molecule. The advantage of polynuclear Ln(III) complexes is that different Ln(III) ions can be used in a single molecule, allowing multiple emission bands [61, 62]. Moreover, when there are two or more Ln(III) ions, ET between Ln(III) ions (LnLnET) occurs depending on the distance between them

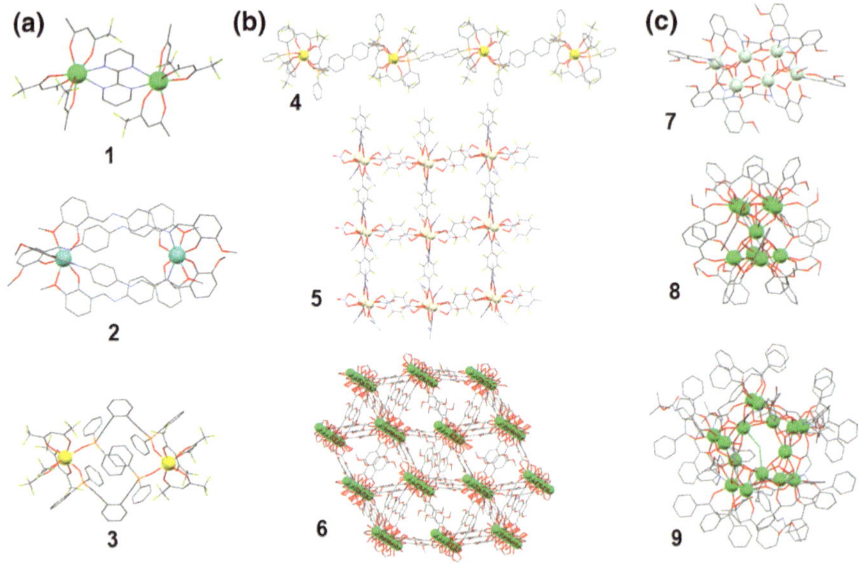


**Fig. 1.9** Examples of **a** polynuclear Ln(III) complexes, **b** Ln(III) coordination polymers, and **c** Ln(III) clusters. **1**: Ref. [64], **2**: Ref. [65], **3**: Ref. [66], **4**: Ref. [69], **5**: Ref. [70], **6**: Ref. [71], **7**: Ref. [80], **8**: Ref. [81], and **9**: Ref. [82]

[61, 63]. Figure 1.9 shows several of the reported polynuclear Ln(III) complexes [64–66].

In general, ET is a distance-dependent process. Förster mechanism and Dexter mechanism are the two widely known mechanisms of the ET process, each corresponding to the direct Coulomb interaction and exchange interaction, respectively, of the approximated electrostatic coupling matrix element [67, 68]. Both are distance dependent, but Förster mechanism [68] has a distance dependence of $R^{-6}$ while Dexter mechanism [67] has an exp($2R/L$) dependence ($L$ is the effective Bohr radius).

Ln(III) coordination polymers are polynuclear Ln(III) complexes with polymer structure (Fig. 1.9b) [69–71]. They are characterized by chain of Ln(III) ions and bridging ligands between them. Porous coordination polymers (PCP), also known as metal-organic frameworks (MOF, Fig. 1.9b bottom), are coordination polymers with large spatial vacancy between the chains [71]. Depending on the ligand design, some of these coordination polymers have high thermostability [70, 72, 73]. MOF, in particular, can be used as an adsorbing agent for their porous structure [74, 75]. Regarding LnLnET, numerous reports are available that utilize temperature dependent LnLnET for thermosensing functions [76–79].

Ln(III) clusters are polynuclear complexes with Ln(III) rich core and multiple ligands coordinating around the core (Fig. 1.9c) [80–82]. The Ln(III) ions in the core are often bridged by $OH^-$ ions and therefore have very short Ln(III)-Ln(III) distance. It is not easy to “design” a synthetic procedure to produce Ln(III) clusters, and their formation remains serendipitous [83–85]. The numbers of the reports on

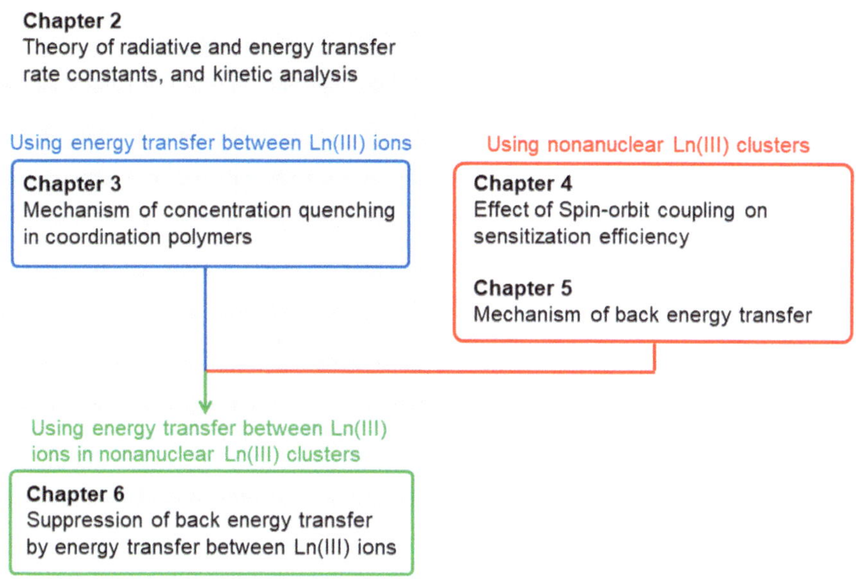


**Fig. 1.10** The outline of this thesis

Ln(III) clusters are limited compared to those of other class of polynuclear Ln(III) complexes likely because of the unpredictable nature of the synthesis of the clusters. Reports regarding luminescent properties of Ln(III) clusters are further limited [82, 86–88]. Structurally, Ln(III) clusters share the traits of both coordination polymers and mononuclear complexes; they involve LnLnET but with a discrete structure (such as size, shape, and molecular weight). Considering the close Ln(III)-Ln(III) distance, LnLnET is expected to be very fast compared to the other polynuclear complexes and could lead to new properties that cannot be achieved in mononuclear complexes or coordination polymers. The structural and luminescent properties of clusters potentially provide a new aspect in Ln(III) complexes.

## 1.4 Outline

The objective of the study is to elucidate various unexplored photophysical processes that could affect the photoluminescence efficiency with emphasis on energy transfer processes between the ligand and Ln(III) ion or between Ln(III) ions. Ln(III) coordination polymers and Ln(III) clusters are the central part of the study. This thesis provides new insight into strategies of molecular design for highly efficient Ln(III) complexes. The outline of this thesis is summarized in Fig. 1.10.

In order to analyze the photophysics of a complicated system of coordination polymers and clusters, quantum mechanical analysis and kinetic analysis are employed. In Chap. 2, the theories that are relevant to the discussion in the following chapters are discussed. The chapter covers the (quantum mechanical) theory of the transition and energy transfer of Ln(III) ions, as well as derivation of rate equations that model the excited state dynamics. The former describes the photophysical property of individual Ln(III) ion, whereas the latter describes how the interaction of these Ln(III) ions through energy transfer results in the actual observed photophysical properties. These two theories are useful in describing complicated photophysical processes where microscopic and macroscopic effects are needed to be accounted for.

Concentration quenching is considered to be an inevitable effect when Ln-to-Ln energy transfer (LnLnET) occurs in Ln(III) doped ceramics [89, 90]. In Ln(III) complexes, this is experimentally considered as a small effect, although their mechanism is not known [76, 91, 92]. In Chap. 3, the concentration quenching mechanism in Yb(III) coordination polymer (in which this host is already reported to show LnLnET) is described both experimentally and theoretically. The theoretical results qualitatively match well with the experimental results when considered with the phonon-assisted energy transfer relaxation model [93], a type of concentration quenching observed only for high-purity solids.

Strong spin-orbit coupling promotes intersystem crossing from the singlet excited state ($S_1$) to the triplet excited state ($T_1$) but at the same time promotes relaxation from $T_1$ state to the ground state ($S_0$). The latter is an unfavorable effect (Eq. 1.6) but is experimentally difficult to confirm because spin-orbit coupling is intrinsic for each Ln(III) ion, and changing Ln(III) ions changes all the other processes unrelated to spin-orbit coupling as well. Ln(III) clusters make this investigation possible because Ln(III) ions can be mixed. Chapter 4 describes the effect of spin-orbit coupling induced by the Gd(III), Yb(III), or Lu(III) ions on the photophysical processes within the organic ligands and their effect on the Yb sensitization efficiency ($\eta_{sens}$) using nonanuclear Yb/Gd or Yb/Lu mixed clusters.

Regarding ligand-Ln energy transfer, back energy transfer (BET) is the reverse process of forward energy transfer (FET) that reduces the overall sensitization efficiency $\eta_{sens}$. In Chap. 5, the effect of the excited state energy of the ligands on the FET and BET rate constants in nonanuclear Tb(III) clusters is described. The factors that affect BET rate constant is derived. This chapter also describes the rigidity of the core structure.

Tb-to-Tb energy transfer (TbTbET) is a competitive process with BET since they both occur from the excited state of a Tb(III) ion. In the case of Tb(III) clusters, the distance between Tb(III) ions is short, and fast TbTbET can be expected. Chapter 6 describes the effect of Tb-to-Tb energy transfer (TbTbET) on the photophysical properties of nonanuclear Tb(III) clusters in the presence of BET.

Chapter 7 summarizes the study described from in this Chapter to Chap. 6 along with a new insight into the strategy for molecular design to achieve Ln(III) complexes with high luminescence efficiency. The chapter is concluded with an outlook.

# References


1. Bragg, W (1935) The universe of light. The Hegeler Institute
2. Imada H, Miwa K, Imai-Imada M, Kawahara S, Kimura K, Kim Y (2016) Real-space investigation of energy transfer in heterogeneous molecular dimers. Nature 538:364–367
3. Di Bartolo B, Goldberg V (1980) Radiationless processes. Plenum Press
4. Kasha M (1950) Characterization of electronic transitions in complex molecules. Discuss Faraday Soc 9:14–19
5. Huang C (2010) Rare earth coordination chemistry: fundamentals and applications. Wiley
6. Bünzli JG, Eliseeva SV (2011) Basics of Lanthanide Photophysics. In: Springer Series on Fluorescence
7. Blackburn OA, Chilton NF, Keller K, Tait CE, Myers WK, McInnes EJL, Kenwright AM, Beer PD, Timmel CR, Faulkner S (2015) Spectroscopic and crystal field consequences of fluoride binding by [Yb· DTMA] 3+ in aqueous solution. Angew Chem Int Ed 54:10783–10786
8. Liu G, Jacquier B (2005) Spectroscopic properties of rare earths in optical materials. Springer
9. Carnall WT, Fields PR, Rajnak K (1968) Electronic energy levels in the trivalent lanthanide aquo ions. I. Pr3+, Nd3+, Pm3+, Sm3+, Dy3+, Ho3+, Er3+, and Tm3+. J Chem Phys 49:4424–4442
10. Carnall WT, Fields PR, Rajnak K (1968) Electronic energy levels of the trivalent lanthanide aquo ions. II. Gd3+. J Chem Phys 49:4443–4446
11. Carnall WT, Fields PR, Rajnak K (1968) Electronic energy levels in the trivalent lanthanide aquo ions. I. Pr3+, Nd3+, Pm3+, Sm3+, Dy3+, Ho3+, Er3+, and Tm3+. J Chem Phys 49:4447–4449
12. Carnall WT, Fields PR, Rajnak K (1968) Electronic energy levels in the trivalent lanthanide aquo ions. I. Pr3+, Nd3+, Pm3+, Sm3+, Dy3+, Ho3+, Er3+, and Tm3+. J Chem Phys 49:4450–4455
13. Dieke GH, Crosswhite HM (1963) The spectra of the doubly and triply ionized rare earths. Appl Opt 2:675–686
14. Carnall WT, Fields PR, Wybourne BG (1965) Spectral intensities of the trivalent lanthanides and actinides in solution. I. Pr3+, Nd3+, Er3+, Tm3+, and Yb3+. J Chem Phys 42:3797–3806
15. Carnall WT, Fields PR, Rajnak K (1968) Spectral intensities of the trivalent lanthanides and actinides in solution. II. Pm3+, Sm3+, Eu3+, Gd3+, Tb3+, Dy3+, and Ho3+. J Chem Phys 49:4412–4423
16. Marcantonatos MD (1986) Multiphonon non-radiative relaxation rates and Judd–Ofelt parameters of lanthanide ions in various solid hosts. J Chem Soc Faraday Trans 2(82):381–393
17. Pukhov KK, Basiev TT, Auzel F, Pellé F, Heber J (2001) Multiphonon sideband intensities in rare earth ions in crystal. J Lumin 94–95:737–741
18. Miyakawa T, Dexter DL (1970) Phonon sidebands, multiphonon relaxation of excited states, and phonon-assisted energy transfer between ions in solids. Phys Rev B 1:2961–2969
19. Auzel F (2004) Upconversion and anti-stokes processes with f and d ions in solids. Chem Rev 104:139–173
20. Binnemans K (2015) Interpretation of europium (III) spectra. Coord Chem Rev 295:1–45
21. Lima NBD, Gonçalves SMC, Júnior SA, Simas AM (2013) A comprehensive strategy to boost the quantum yield of luminescence of europium complexes. Sci Rep 3:2395
22. Yanagisawa K, Kitagawa Y, Nakanishi T, Akama T, Kobayashi M, Seki T, Hasegawa Y (2017) Enhanced luminescence of asymmetrical seven-coordinate EuIII complexes including LMCT perturbation. Euro J Inorganic Chem 2017(32):3843–3848
23. Judd BR (1962) Optical absorption intensities of rare-earth ions. Phys Rev 127:750–761
24. Ofelt GS (1962) Intensities of crystal spectra of rare-earth ions. J Chem Phys 37:511–520
25. Hehlen MP, Brik MG, Krämer KW (2013) 50th anniversary of the Judd–Ofelt theory: An experimentalist's view of the formalism and its application. J Lumin 136:221–239
26. Bünzli J-CG, Eliseeva SV (2013) Intriguing aspects of lanthanide luminescence. Chem Sci 4:1939–1949
27. Hasegawa Y, Wada Y, Yanagida S (2004) Strategies for the design of luminescent lanthanide (III) complexes and their photonic applications. J Photochem Photobiol C 5:183–202
28. Bünzli J-CG (2010) Lanthanide luminescence for biomedical analyses and imaging. Chem Rev 110:2729–2755

29. Bünzli J-CG, Piguet C (2005) Taking advantage of luminescent lanthanide ions. Chem Soc Rev 34:1048–1077
30. Eliseeva SV, Bünzli J-CG (2011) Rare earths: jewels for functional materials of the future. New J Chem 35:1165
31. Rodríguez-Rodríguez A, Esteban-Gómez D, Tripier R, Tircsó G, Garda Z, Tóth I, de Blas A, Rodríguez-Blas T, Platas-Iglesias C (2014) A. deBlas, T. Rodriguez-Blas, C. Platas-Iglesias. J Am Chem Soc 136:17954–17957
32. Souri N, Tian P, Lecointre A, Lemaire Z, Chafaa S, Strub J-M, Cianférani S, Elhabiri M, Platas-Iglesias C, Charbonnière LJ (2016) Step by Step Assembly of Polynuclear Lanthanide Complexes with a Phosphonated Bipyridine Ligand. Inorg Chem 55:12962–12974
33. Chauvin A-S, Comby S, Song B, Vandevyver CDB, Bünzli J-CG (2008) A versatile ditopic ligand system for sensitizing the luminescence of bimetallic lanthanide bio-imaging probes. Chem Eur J 14:1726–1739
34. Eliseeva SV, Bünzli J-CG (2010) Lanthanide luminescence for functional materials and bio-sciences. Chem Soc Rev 39:189–227
35. Tsien RY, Ernst L, Waggoner A (2006) Fluorophores for confocal microscopy: photophysics and photochemistry. In: Handbook of biological confocal microscopy. Springer
36. Placide V, Bui AT, Grichine A, Duperray A, Pitrat D, Andraud C, Maury O (2015) wo-photon multiplexing bio-imaging using a combination of Eu-and Tb-bioprobes. Dalton Trans 44:4918–4924
37. Nakamura K, Hasegawa Y, Kawai H, Yasuda N, Kanehisa N, Kai Y, Nagamura T, Yanagida S, Wada Y (2007) Enhanced lasing properties of dissymmetric Eu (III) complex with bidentate phosphine ligands. J Phys Chem A 111:3029–3037
38. Marchionna S, Meinardi F, Acciarri M, Binetti S, Papagni A, Pizzini S, Malatesta V, Tubino R (2006) Photovoltaic quantum efficiency enhancement by light harvesting of organo-lanthanide complexes. J Lumin 118:325–329
39. Gunnlaugsson T, Leonard JP (2005) Responsive lanthanide luminescent cyclen complexes: from switching/sensing to supramolecular architectures. Chem Commun 25:3114–3131
40. Bradberry SJ, Savyasachi AJ, Martinez-Calvo M, Gunnlaugsson T (2014) Development of responsive visibly and nir luminescent and supramolecular coordination self-assemblies using lanthanide ion directed synthesis. Coord Chem Rev 273–274:226–241
41. Taydakov IV, Akkuzina AA, Avetisov RI, Khomyakov AV, Saifutyarov RR, Avetissov IC (2016) Effective electroluminescent materials for OLED applications based on lanthanide 1.3-diketonates bearing pyrazole moiety. J Lumin 177:31–39
42. Kido J, Okamoto Y (2002) Organo lanthanide metal complexes for electroluminescent materials. Chem Rev 102:2357–2368
43. de Bettencourt-Dias A (2007) Lanthanide-based emitting materials in light-emitting diodes. Dalton Trans 22:2229–2241
44. Shahi PK, Singh AK, Rai SB, Ullrich B (2015) Lanthanide complexes for temperature sensing, UV light detection, and laser applications. Sens Actuators A 222:255–261
45. Teixeira ES, Neto BB, de Oliveira PMC, Longo RL (2016) Chemometric analysis of the luminescence quantum yields in lanthanide ion complexes. J Lumin 170:602–613
46. Beeby A, Clarkson IM, Dickins RS, Faulkner S, Parker D, Royle L, de Sousa AS, Williams JAG, Woods M (1999) Non-radiative deactivation of the excited states of europium, terbium and ytterbium complexes by proximate energy-matched OH, NH and CH oscillators: an improved luminescence method for establishing solution hydration states. J Chem Soc Perkin Trans 2:493–504
47. Hasegawa Y, Kimura Y, Murakoshi K, Wada Y, Kim J-H, Nakashima N, Yamanaka T, Yanagida S (1996) Enhanced emission of deuterated tris (hexafluoroacetylacetonato) neodymium (III) complex in solution by suppression of radiationless transition via vibrational excitation. J Phys Chem 100:10201–10205
48. Hasegawa Y, Murakoshi K, Wada Y, Yanagida S, Kim J-H, Nakashima N, Yamanaka T (1996) Enhancement of luminescence of Nd3+ complexes with deuterated hexafluoroacetylacetonato ligands in organic solvent. Chem Phys Lett 248:8–12

49. Doffek C, Alzakhem N, Bischof C, Wahsner J, Güden-Silber T, Lügger J, Platas-Iglesias C, Seitz M (2012) Understanding the quenching effects of aromatic C–H-and C–D-oscillators in near-IR lanthanoid luminescence. J Am Chem Soc 134:16413–16423
50. Andres J, Chauvin A-S (2013) Energy transfer in coumarin-sensitised lanthanide luminescence: investigation of the nature of the sensitiser and its distance to the lanthanide ion. Phys Chem Chem Phys 15:15981–15994
51. Albuquerque RQ, Da Costa NB, Freire RO (2011) Design of new highly luminescent Tb3+ complexes using theoretical combinatorial chemistry. J Lumin 131:2487–2491
52. Gutierrez F, Tedeschi C, Maron L, Daudey JP, Poteau R, Azema J, Tisnès C, Picard C (2004) Quantum chemistry-based interpretations on the lowest triplet state of luminescent lanthanides complexes. Part 1. Relation between the triplet state energy of hydroxamate complexes and their luminescence properties. Dalton Trans 9:1334–1347
53. Gutierrez F, Tedeschi C, Maron L, Daudey J-P, Azema J, Tisnès P, Picard C, Poteau R (2005) Quantum chemistry-based interpretations on the lowest triplet state of luminescent lanthanides complexes. Part 2. Influence of the electrostatic interactions on the triplet state energy of terbium complexes. J Mol Struct (Theochem) 756:151–162
54. Botelho MB, Gálvez-López MD, De Cola L, Albuquerque RQ, de Camargo AS (2013) Towards the design of highly luminescent europium (III) complexes. Euro J Inorganic Chem 2013(29):5064–5070
55. Latva M, Takalo H, Mukkala V-M, Matachescu C, Rodríguez-Ubis JC, Kankare J (1997) Correlation between the lowest triplet state energy level of the ligand and lanthanide (III) luminescence quantum yield. J Lumin 75:149–169
56. Ha-Thi M-H, Delaire JA, Michelet V, Leray I (2010) Sensitized emission of luminescent lanthanide complexes based on a phosphane oxide derivative. J Phys Chem A 114:3264–3269
57. Fu L-M, Ai X-C, Li MY, Wen XF, Hao R, Wu Y-S, Wang Y, Zhang J-P (2010) Role of ligand-to-metal charge transfer state in nontriplet photosensitization of luminescent europium complex. J Phys Chem A 114:4494–4500
58. Tobita S, Arakawa M, Tanaka I (1985) The paramagnetic metal effect on the ligand localized S1. apprx. fwdarw. T1 intersystem crossing in the rare-earth-metal complexes with methyl salicylate. J Phys Chem 89:5649–5654
59. Guldi DM, Mody TD, Gerasimchuk NN, Magda D, Sessler JL (2000) Influence of large metal cations on the photophysical properties of texaphyrin, a rigid aromatic chromophore. J A Chem Soc 122:8289–8298
60. Marian CM (2012) Spin–orbit coupling and intersystem crossing in molecules. WIREs Comput Mol Sci 2:187–203
61. Lewis DJ, Glover PB, Solomons MC, Pikramenou Z (2011) Purely heterometallic lanthanide (III) macrocycles through controlled assembly of disulfide bonds for dual color emission. J Am Chem Soc 133:1033–1043
62. D'Aléo A, Pointillart F, Ouahab L, Andraud C, Maury O (2012) Charge transfer excited states sensitization of lanthanide emitting from the visible to the near-infra-red. Coord Chem Rev 256:1604–1620
63. Zaïm A, Eliseeva SV, Guénée L, Nozary H, Petoud S, Piguet C (2014) Lanthanide-to-lanthanide energy-transfer processes operating in discrete polynuclear complexes: can trivalent europium be used as a local structural probe? Chem Eur J 20:12172–12182
64. Ilmi R, Iftikhar K (2015) Optical emission studies of new europium and terbium dinuclear complexes with trifluoroacetylacetone and bridging bipyrimidine. Fast radiation and high emission quantum yield. Polyhedron 102:16–26
65. Habib F, Long J, Lin P-H, Korobkov I, Ungur L, Wernsdorfer W, Chibotaru LF, Murugesu M (2012) Supramolecular architectures for controlling slow magnetic relaxation in field-induced single-molecule magnets. Chem Sci 3:2158
66. Hirai Y, Nakanishi T, Kitagawa Y, Fushimi K, Seki T, Ito H, Fueno H, Tanaka K, Satoh T, Hasegawa Y (2015) Luminescent coordination glass: remarkable morphological strategy for assembled Eu (III) complexes. Inorg Chem 54:4364–4370
67. Dexter DL (1953) A theory of sensitized luminescence in solids. J Chem Phys 21:836–850

68. Förster T (1959)10th Spiers memorial lecture. Transfer mechanisms of electronic excitation. Disc Faraday Soc 27:7–17
69. Seidel C, Lorbeer C, Cybińska J, Mudring A-V, Ruschewitz U (2012) Lanthanide coordination polymers with tetrafluoroterephthalate as a bridging ligand: thermal and optical properties. Inorg Chem 51:4679–4688
70. Miyata K, Ohba T, Kobayashi A, Kato M, Nakanishi T, Fushimi K, Hasegawa Y (2012) Thermostable Organo-phosphor: Low-Vibrational Coordination Polymers That Exhibit Different Intermolecular Interactions. Chem Plus Chem 77:277–280
71. Cui Y, Xu H, Yue Y, Guo Z, Yu J, Chen Z, Gao J, Yang Y, Qian G, Chen B (2012) A luminescent mixed-lanthanide metal–organic framework thermometer. J Am Chem Soc 134:3979–3982
72. Nakajima A, Nakanishi T, Kitagawa Y, Seki T, Ito H, Fushimi K, Hasegawa Y (2016) Hyper-stable organo-Eu III luminophore under high temperature for photo-industrial application. Sci Rep 6:24458
73. Zhang H, Zhou L, Wei J, Li Z, Lin P, Du S (2012) Highly luminescent and thermostable lanthanide-carboxylate framework materials with helical configurations. J Mater Chem 22:21210–21217
74. Zhu Y, Zhu M, Xia L, Wu Y, Hua H, Xie J (2016) Lanthanide metal-organic frameworks with six-coordinated ln (iii) ions and free functional organic sites for adsorptions and extensive catalytic activities. Sci Rep 6:29728
75. Lu W-G, Zhong D-C, Jiang L, Lu T-B (2012) Lanthanide coordination polymers constructed from Imidazole-4, 5-dicarboxylate and sulfate: syntheses, structural diversity, and photoluminescent properties. Cryst Growth Des 12:3675–3683
76. Cui Y, Chen B, Qian G (2014) Lanthanide metal-organic frameworks for luminescent sensing and light-emitting applications. Coord Chem Rev 273–274:76–86
77. Cui Y, Song R, Yu J, Liu M, Wang Z, Wu C, Yang Y, Wang Z, Chen B, Qian G (2015) Dual-emitting MOF⊃ dye composite for ratiometric temperature sensing. Adv Mater 27:1420–1425
78. Hirai Y, Nakanishi T, Miyata K, Fushimi K, Hasegawa Y (2014) Thermo-sensitive luminescent materials composed of Tb (III) and Eu (III) complexes. Mater Lett 130:91–93
79. Rocha J, Brites CDS, Carlos LD (2016) Lanthanide organic framework luminescent thermometers. Chem Eur J 22:14782–14795
80. Liao S, Yang X, Jones RA (2012) Self-assembly of luminescent hexanuclear lanthanide salen complexes. Cryst Growth Des 12:970–974
81. Nakanishi T, Suzuki Y, Doi Y, Seki T, Koizumi H, Fushimi K, Fujita K, Hinatsu Y, Ito H, Tanaka K, Hasegawa Y (2014) Enhancement of optical faraday effect of nonanuclear Tb (III) complexes. Inorg Chem 53:7635–7641
82. Thielemann DT, Wagner AT, Rösch E, Kölmel DK, Heck JG, Rudat B, Neumaier M, Feldmann C, Schepers U, Bräse S, Roesky PW (2013) Luminescent cell-penetrating pentadecanuclear lanthanide clusters. J Am Chem Soc 135:7454–7457
83. Petit S, Baril-Robert F, Pilet G, Reber C, Luneau D (2009) Luminescence spectroscopy of europium (III) and terbium (III) penta-, octa-and nonanuclear clusters with β-diketonate ligands. Dalton Trans 34:6809–6815
84. Wang R, Song D, Wang S (2002) Ligand-controlled self-assembly of polynuclear lanthanide–oxo/hydroxo complexes: from synthetic serendipity to rational supramolecular design. Chem Commun 24:368–369
85. Zheng Z (2001) Ligand-controlled self-assembly of polynuclear lanthanide–oxo/hydroxo complexes: from synthetic serendipity to rational supramolecular design. Chem Commun 24:2521–2529
86. Kornienko A, Emge TJ, Kumar GA, Riman RE, Brennan JG (2005) Lanthanide clusters with internal Ln ions: Highly emissive molecules with solid-state cores. J Am Chem Soc 127:3501–3505
87. Alexandropoulos DI, Mukherjee S, Papatriantafyllopoulou C, Raptopoulou CP, Psycharis V, Bekiari V, Christou G, Stamatatos TC (2011) A new family of nonanuclear lanthanide clusters displaying magnetic and optical properties. Inorg Chem 50:11276–11278

88. Mazarakioti EC, Poole KM, Cunha-Silva L, Christou G, Stamatatos TC (2014) A new family of Ln 7 clusters with an ideal D 3h metal-centered trigonal prismatic geometry, and SMM and photoluminescence behaviors. Dalton Trans 43:11456–11460
89. Auzel F, Baldacchini G, Laversenne L, Boulon G (2003) Radiation trapping and self-quenching analysis in Yb3+, Er3+, and Ho3+ doped Y2O3. Opt Mater 24:103–109
90. Auzel F, Pellé F (1996) Concentration and excitation effects in multiphonon non-radiative transitions of rare-earth ions. J Lumin 69:249–255
91. Yan B, Bai Y, Chen Z (2005) Synthesis, structure and luminescence of novel 1D chain coordination polymers [Ln (isophth)(Hisophth)(H 2 O) 4· 4H 2 O] n (Ln= Sm, Dy). J Mol Struct 741:141–147
92. Heine J, Müller-Buschbaum K (2013) Engineering metal-based luminescence in coordination polymers and metal–organic frameworks. Chem Soc Rev 43:9232–9242
93. Auzel F (2002) A fundamental self-generated quenching center for lanthanide-doped high-purity solids. J Lumin 100:125–130

# Chapter 2
# Theory

**Abstract** The upcoming chapters require that the readers have some background in the theory of lanthanide luminescence and the theory of kinetic analysis. The theory of lanthanide luminescence, or the infamous Judd-Ofelt theory, is one of the most critical development in lanthanide physics and chemistry that allowed great advancement in the understanding of lanthanide spectroscopy. Kinetic analysis allows one to model and calculate the dynamics of an interacting system. The powerful method of combining these two theories allows one to simulate the excited state dynamics of interacting multi-lanthanide system that would otherwise be too complicated to discuss. This chapter describes the basic concept of these two theories as well as the implication of the resulting equations.

## 2.1 Introduction

Photophysical properties of a material can be characterized by emission and absorption spectroscopy, emission lifetime measurements, and emission quantum yield measurements. These spectroscopic methods measure photophysical properties of an ensemble of multiple emission centers within a material. The photophysical properties of a single emission center can be measured, provided that the measurement apparatus is observing an ideal system, i.e., the emission centers are homogeneous and non-interacting. In a system with inhomogeneous and interacting emission centers, the obtained data requires statistically rigorous treatment. Many of the photophysical properties of lanthanide complexes described in this thesis result from the emission from an ensemble of species that are interacting with each other. For instance, Chap. 3 describes Yb(III) ions that are distributed inhomogeneously throughout the coordination polymer host. Chapter 6 discusses the energy transfer between Tb(III) ions as well as the energy transfer of Tb(III) ions to and from the organic ligands. Although the details will be discussed in their respective chapters, these inhomogeneous and interacting system complicates the interpretation of the data.

 
S. Omagari, *Energy Transfer Processes in Polynuclear Lanthanide Complexes*, Springer Theses, https://doi.org/10.1007/978-981-13-6049-7_2

The theory that models the dynamics of a system, in general, comprises of microscopic and macroscopic theories that are independent of each other. To a microscopic scale, each species is emitting a photon, nonradiatively relaxing to ground state, and/or transferring its energy to other species. Such properties can be calculated quantum mechanically. In Sect. 2.2, the theories of radiative transitions and energy transfer in Ln(III) ions are reviewed. The theory of radiative transitions for Ln(III) ions is widely known as the Judd-Ofelt theory [1, 2]. The theory of energy transfer was provided by T. Kushida in 1973 under the framework of the Judd-Ofelt theory [3]. In Sect. 2.3, the theory for modeling and deriving the rate equation of photophysical properties of a system of interacting species is reviewed [4]. A simple model of mononuclear Ln(III) complex is provided as an example.

## 2.2 Lanthanides

As mentioned in Chap. 1, the radiative transition and nonradiative energy transfer of Ln(III) ions have been extensively investigated in the early 1900s. The breakthrough in understanding the photophysical properties of Ln(III) ions came after B. R. Judd and G. S. Ofelt who separately reported a semi-empirical, quantum mechanical theory in 1962 [1, 2]. The theory is commonly referred to as the "Judd-Ofelt theory" [5], and it is the framework for the theory of energy transfer between Ln(III) ions by Kushida [3] and the theory that was later revisited by Malta [6]. The following subsections describe the essential part of the theory with emphasis on the applicative aspect as they will become important in the following chapters. The rigorous derivation of the equations is beyond the scope of this work and will be kept to a minimum.

The calculation of radiative transition rate constant in Judd-Ofelt theory is a multistep calculation. The basic procedure is as follows [5]:

(1) Calculation of reduced matrix elements for various tensor operators arising from transition dipole moment operator.
(2) Fit the Slater radial integrals and spin-orbit coupling parameters to the experimental 4f-state energy levels to obtain the "mixed" 4f-state wavefunctions.
(3) Calculation of theoretical oscillator strengths for electric dipole and magnetic dipole transitions between 4f-states using the "mixed" 4f-state wavefunctions.
(4) Fit the "Judd-Ofelt" parameters (which includes 4f5d-state mixed by crystal field perturbation) to the experimental 4f-state transition oscillator strengths to obtain the transition rate constants.

The calculation of energy transfer rate constant in the study reported by T. Kushida and O. L. Malta is based on the Judd-Ofelt theory. As such, most of the parameters needed for the calculation can be directly taken from the parameters obtained in the above procedure. The focus here is the resonance mechanism, and the exchange mechanism is neglected. The details of the entire procedure are explained below.

### 2.2.1 Energy Levels and Wavefunctions

The 4f-electronic states in Judd-Ofelt theory are treated as a hydrogen-like atom under a central-field approximation. The intermediate coupling (electrostatic and spin-orbit interactions in a fairly equal contribution) can then be considered to provide a first-order description to the wavefunction and the energy levels of 4f-states [7]. Finally, the effect of crystal-field is treated once again as a first-order perturbation in which experimental parameters are used [5]. The theory of the energy levels and wavefunctions is based on the tensor operator technique since spherical harmonics can best simplify calculations for hydrogen-like atoms. Similar technique was used for the infamous Tanabe-Sugano diagram for d-electron metals expect electrostatic and crystal-field gives the first-order description to the wavefunction and the energy levels of the d-states, and the effect of spin-orbit coupling is considered as the first-order perturbation.

First, the quantum numbers relevant to the initial wavefunctions (the "pure" 4f-states) are reviewed. Since the electrons only in the 4f-orbitals are being considered, the principal quantum number is $n = 4$ and the orbital angular momentum (a.k.a. azimuthal) quantum number is $l = 3$. Electrons also have intrinsic angular momentum called spin with electron spin (projection) quantum number $m_s$. For a given $4f^N$ electron configuration ($N$ being the number of electrons), there are $N$ electron spins each with angular momentum $s_i$ (quantum number $m_s = \pm 1/2$) and $N$ 4f-orbitals each with angular momentum $l_i$ (quantum number $m_l = -3, \ldots, 3$), where $i = 1, 2, \ldots, N$. These two angular momenta are combined to give a total spin angular momentum $S = \sum_i s_i$ and total orbital angular momentum $L = \sum_i l_i$. Each of these momentum has quantum numbers of $M_s = -S, \ldots, S$ and $M_l = -L, \ldots, L$, respectively. The combination must be within the Pauli's exclusion principle, so for $N$ 4f-electrons, there are $\frac{14!}{N!(14-N)!}$ possible combinations. At this point, the initial quantum numbers $m_s$ and $m_l$ can no longer represent each state and instead $S$ and $L$ become good quantum numbers. It should be noted that some lanthanide $4f^N$-states have two or more states of the same $S$ and $L$. These two states are distinguished by seniority quantum number $\alpha$. The splitting of the zeroth-order degenerate $4f^N$-states into these new states is caused by electrostatic coupling between electrons. Furthermore, spin-orbit coupling causes magnetic moment arising from the electron spin to be coupled with the magnetic moment arising from the electron's orbit. The resulting total angular momentum quantum number $J$ becomes a good quantum number in this coupling. In the $LS$-coupling scheme (in contrast to the $jj$-coupling scheme), the states produced by the electrostatic coupling are further split into $J = |L - S|, |L - S + 1|, \ldots, |L + S|$. These states are described using the Russel-Saunders term symbol $^{2S+1}L_J$. Finding all of the combination of angular momenta for all lanthanides shows that a lanthanide with $N$ electrons will have same ${}^{2S+1}L_J$ term symbols as that with 14-$N$ electrons. The term "intermediate coupling" refers to a case where the electrostatic coupling and spin-orbit coupling are approximately in the same order of energy [5].

Using the wavefunctions of "pure" Russel-Saunders 4f-states as a basis, the "mixed" 4f-states under the intermediate coupling scheme are calculated. The Hamiltonian for the intermediate coupling scheme is

$$H = H_{\mathrm{ee}} + H_{\mathrm{SO}}, \tag{2.1}$$

where $H_{\mathrm{ee}}$ is the electrostatic coupling Hamiltonian and $H_{\mathrm{SO}}$ is the spin-orbit coupling Hamiltonian. Since a hydrogen-like atom is being considered, the wavefunction $\Psi(r, \theta, \emptyset)$ is a product of radial wavefunction $R_{n,l}(r)$ and spherical harmonics $Y_{l,m}(\theta, \emptyset)$, so that $\Psi(r, \theta, \emptyset) = R_{n,l}(r)Y_{l,m}(\theta, \emptyset)$. For spherical harmonics, tensor operator technique can be utilized. Electrostatic coupling Hamiltonian for 4f-electrons is given by:

$$H_{\mathrm{ee}} = \sum_{i \neq j} \frac{e^2}{4\pi \varepsilon_0} \frac{1}{r_{ij}} = \frac{1}{4\pi \varepsilon_0} \sum_{i \neq j} \sum_{k=0,2,4,6} C_i^{(k)} C_j^{(k)}, \tag{2.2}$$

where $e$, $\varepsilon_0$, and $r_{ij}$ are elementary charge, vacuum permittivity, and the distance between electron $i$ and $j$. $C_q^{(k)} = \left(\frac{4\pi}{2k+1}\right)^{\frac{1}{2}} Y_{k,q}$ is a tensor operator with spherical harmonics $Y_{k,q}$. The matrix element of $H_{\mathrm{ee}}$ is expressed for 4f-electrons as:

$$\begin{aligned} \left\langle l^N SL \middle| C_i^{(k)} C_j^{(k)} \middle| l^N S'L' \right\rangle &= \delta_{SS'} \delta_{LL'} (-1)^{4\times 3} (-1)^L (2L+1)^2 \\ &\times \sum_{k=0,2,4,6} \begin{pmatrix} 3 & k & 3 \\ 0 & 0 & 0 \end{pmatrix}^2 \begin{Bmatrix} 3 & 3 & k \\ 3 & 3 & L \end{Bmatrix} F_{(k)}, \end{aligned} \tag{2.3}$$

where $\begin{pmatrix} a & b & c \\ d & e & f \end{pmatrix}$ and $\begin{Bmatrix} a & b & c \\ d & e & f \end{Bmatrix}$ are the Wigner 3-j symbol and Wigner 6-j symbol, respectively. These coefficients arise from the coupling of angular momentum, and allow great simplification of the calculation of the matrix elements. $F_{(k)}$ is the Slater integrals. The Dirac's delta arises from the fact that $H_{\mathrm{ee}}$ does not act on a spin. In order to prevent off-diagonal elements due to some lanthanides having multiple 4f$^N$-states of the same $S$ and $L$, C. W. Nielson and G. F. Koster have provided matrix elements in the form [8]:

$$\left\langle l^N SL | H_{\mathrm{ee}} | l^N S'L' \right\rangle = \sum_{p=0.1.2.3} e_p E^{(p)} \tag{2.4}$$

where $e_p$ are coefficients listed in the book by the two authors, and $E^{(p)}$ are expressed with Slater integrals as:

$$E^{(0)} = F_{(0)} - 10F_{(2)} - 33F_{(4)} - 286F_{(6)}, \tag{2.5a}$$

$$E^{(1)} = \frac{1}{9}\left(70F_{(2)} + 231F_{(4)} + 2002F_{(6)}\right), \tag{2.5b}$$

$$E^{(2)} = \frac{1}{9}\left(F_{(2)} - 3F_{(4)} + 7F_{(6)}\right), \tag{2.5c}$$

$$E^{(3)} = \frac{1}{3}\left(5F_{(2)} + 6F_{(4)} - 91F_{(6)}\right). \tag{2.5d}$$

$F_{(0)}$ is usually considered to be zero since it will only shift all of the energies in uniform and does not provide any meaningful information since the ground state energy is considered to be zero [5].

As for the spin-orbit coupling $H_{SO}$:

$$H_{SO} = \Sigma_i \xi(r_i)\, s_i \cdot l_i = \Sigma_i \frac{\hbar^2}{m_e{}^2 c^2 r_i} \frac{dU(r_i)}{dr_i} s_i \cdot l_i\ , \tag{2.6}$$

where, $\hbar$, $m_e$, and $c$ are Planck constant, electron mass, and the speed of light in vacuum, respectively. $\xi(r_i)$ is the spin-orbit coupling parameter that includes $U(r_i)$, the central-field potential arising from the central-field approximation [7]. The spin-orbit coupling parameter is independent of spin and angular momenta so it is the same for all $4f^N$-states for a given $N$. The matrix element of $H_{SO}$ is expressed in a simplified form under Wigner-Eckart theorem [9] as:

$$\left\langle l^N SLJ \middle| H_{SO} \middle| l^N S'L'J' \right\rangle = \xi(-1)^{J+L+S'} \sqrt{3(3+1)(23+1)} \times \begin{Bmatrix} S & S' & k \\ L' & L & J \end{Bmatrix} \left\langle l^N SL \middle\| V^{(11)} \middle\| l^N S'L \right\rangle. \tag{2.7}$$

where $\left\langle l^N SL \middle\| V^{(11)} \middle\| l^N S'L' \right\rangle$ is the reduced matrix element with tensor operator $V^{(11)}$. Equation (2.5) shows that the matrix element of $H_{SO}$ can be split into two factors: Clebsch-Gordan coefficient [which is directly related to the Wigner 6-j symbol hence their existence in Eq. (2.5)] and the reduced matrix element [5, 7, 8]. The former is a geometrical factor that does not depend on the nature of the tensor operator while the latter depends on the nature of the tensor operator but does not depend on the geometrical factor. This greatly simplifies the calculation. $V^{(11)}$ is given by:

$$\left\langle l^N SL \middle\| V^{(11)} \middle\| l^N S'L' \right\rangle = N\sqrt{\frac{1}{2}\left(\frac{1}{2}+1\right)(2S+1)(2L+1)(2S'+1)(2L'+1)}$$
$$\times \sum\nolimits_{\psi(l^{N-1})} \left(l^{N-1}\bar{S}\bar{L}|l^N SL\right) \begin{Bmatrix} S & S' & 1 \\ \frac{1}{2} & \frac{1}{2} & \bar{S} \end{Bmatrix}$$
$$\begin{Bmatrix} L & L' & 1 \\ 3 & 3 & \bar{L} \end{Bmatrix} (-1)^{\bar{S}+\bar{L}+S+L+\frac{1}{2}+3+1+1}, \tag{2.8}$$

where $\left(l^{N-1}\bar{S}\bar{L}|l^N SL\right)$ is coefficient of fractional parentage and they are summed over all the $4f^{N-1}$-states (4f-states of the previous atomic number). The values are available from the works of Nielson and Koster [8]. It should be noted that the relationship between $4f^N$ and $4f^{14-N}$ lanthanides for the reduced matrix element of tensor operator of $V^{(11)}$ is:

$$\left\langle l^{14-N} SL \middle\| V^{(11)} \middle\| l^{14-N} S'L' \right\rangle = -\left\langle l^N SL \middle\| V^{(11)} \middle\| l^N S'L' \right\rangle. \tag{2.9}$$

$H_{SO}$ contains off-diagonal matrix elements and mixes all states of the same $J$.

Using Eqs. (2.4) and (2.8), the matrix element of Eq. (2.1) can be evaluated. Diagonalization of Eq. (2.1) along with experimental energies of $4f^N$-states derived from absorption spectrum provides an overdetermined system of equations. Nonlinear optimization of the system of equations provides $F_{(2)}$, $F_{(4)}$, $F_{(6)}$, and $\xi$. Using these parameters, the "mixed" wavefucntions of the $4f^N$-states $\left|l^N SLJ_{\text{mixed}}\right\rangle$ can be derived from linear combinations of "pure" wavefunctions $\left|l^N S'L'J_{\text{pure}}\right\rangle$ such that:

$$\left|l^N SLJ_{\text{mixed}}\right\rangle = \sum_i c_i \left|l^N S'L'J_{\text{pure}}\right\rangle. \tag{2.10}$$

The importance of considering electrostatic and spin-orbit coupling is that these coupling sometimes leads to a significant mixture of pure wavefunctions, leading to mixed wavefunctions that do not retain almost 50% of its pure state [5]. It is also worth mentioning that the Slater integrals $F_{(2)}$, $F_{(4)}$, $F_{(6)}$, and spin-orbit coupling parameter $\xi$ are to some degree host-dependent values although they are similar among hosts for a same lanthanide. This means that the optimization of the wavefunctions must be performed for every host material that is being considered.

Several things can be said about the energy levels of lanthanides presented here from the details of the theory. First, looking at the Dieke diagram [10] (Fig. 2.1) in the large energy scale, the energy levels seem symmetric in respect to Gd(III) ion. The number of configurations of 4f-electrons is equal (and so is the term symbols) for lanthanide with $N$ and 14-$N$ electrons, and the number of configurations increases from $N = 1$ to 7 and then decreases from $N = 7$ to 13 (hence Eq. (2.9)).

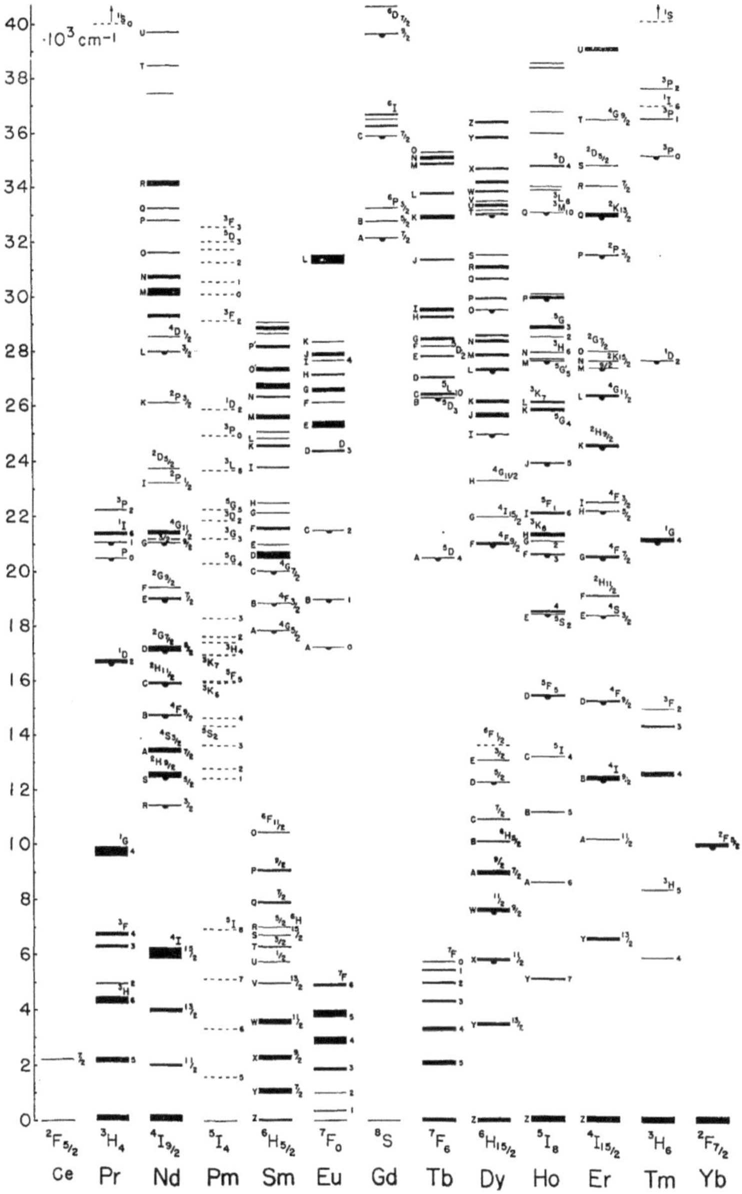


**Fig. 2.1** Dieke diagram representing 4f-state energy levels. Reprinted from Ref. [10] with permission from The Optical Society

Second, looking at the small energy scale, the energy gap between 4f$N$-states of the same $S$ and $L$ is larger with an increase in the number of 4f-electrons. For example, Ce(III) and Yb(III) ions both have the same $^2F_{7/2}$ and $^2F_{5/2}$ terms but Yb(III) ion shows more than twice as large energy gap between these two terms. This is due to the stronger spin-orbit coupling parameter $\xi$ of lanthanide with larger $N$ (heavier atom).

## 2.2.2 *Radiative Rate Constants*

The 4f-4f transition of lanthanide is a parity forbidden transition (also known as Laporte forbidden transition). A transition moment $\vec{\mu}_{if}$ between an initial $\Psi_i$ and a final state $\Psi_f$ is expressed with an electric dipole moment operator $\vec{\mu}$ as $\vec{\mu}_{if} = \int \Psi_i \vec{\mu} \Psi_f d\tau$. The electric dipole moment operator and one electron 4f-orbitals have odd parity. Therefore, the product in the integral also have odd parity if the $4f^N$-wavefunctions of a lanthanide are odd (odd number of 4f-electrons) or even (even number of 4f-electrons). The integration over the product leads to a value of zero. This is the origin of the Laporte forbidden transition of 4f-4f transition. In reality, however, the 4f-4f transitions of lanthanides are observed both in the absorption and emission spectra. The premise of the Judd-Ofelt theory is that the energetically close state, i.e., the $4f^{N-1}5d$-state, is mixed into the $4f^N$-state wavefunctions by crystal field perturbation [1, 2, 5]:

$$\left|l^N SLJ_{\text{actual}}\right\rangle = \left|l^N SLJ_{\text{mixed}}\right\rangle + \sum_{\varphi_{n,l}} \left\{ \frac{\left\langle l^N SLJ_{\text{mixed}} \middle| V \middle| \varphi_{n,l} \right\rangle}{E\left(l^N SLJ_{\text{mixed}}\right) - E\left(\varphi_{n,l}\right)} \left|\varphi_{n,l}\right\rangle \right\}. \tag{2.11}$$

where $\left|l^N SLJ_{\text{actual}}\right\rangle$ and $\left|\varphi_{n,l}\right\rangle$ are wavefunction of $4f^N$-state after mixture of even parity states and wavefunction of the even parity state, respectively. $V$ represents the crystal field interaction operator. With an electric dipole moment operator in tensor operator form, the matrix element of the electric dipole moment is:

$$\left\langle l^N SLJ_{\text{actual}} \middle| D_q^{(1)} \middle| l^N S'L'J'_{\text{actual}} \right\rangle = \sum_{\varphi_{n,l}} \left\{ \frac{\left\langle l^N SLJ_{\text{mixed}} \middle| V \middle| \varphi_{n,l} \right\rangle \left\langle l^N S'L'J'_{\text{mixed}} \middle| D_q^{(1)} \middle| \varphi_{n,l} \right\rangle}{E\left(l^N SLJ\right) - E\left(\varphi_{n,l}\right)} + \frac{\left\langle l^N S'L'J'_{\text{mixed}} \middle| V \middle| \varphi_{n,l} \right\rangle \left\langle l^N SLJ_{\text{mixed}} \middle| D_q^{(1)} \middle| \varphi_{n,l} \right\rangle}{E\left(l^N S'L'J'\right) - E\left(\varphi_{n,l}\right)} \right\} \tag{2.12}$$

In order to simplify Eq. (2.12), two rough assumptions are made: (1) even parity states are all degenerate, and (2) $E\left(l^N SLJ\right) - E\left(\varphi_{n,l}\right) = E\left(l^N S'L'J'\right) - E\left(\varphi_{n,l}\right)$.

With these assumptions, the oscillator strength of electric dipole transition $f_{ED}$ is:

$$f_{ED} = \frac{8\pi^2 m_e}{3h}\frac{v_{abs}}{2J+1}\chi_{abs}\sum_{\lambda=2,4,6}\Omega_\lambda\left|\left\langle l^N SLJ_{mixed}\middle\|U^{(\lambda)}\middle\|l^N S'L'J'_{mixed}\right\rangle\right|^2, \quad (2.13)$$

where $v_{abs}$, $h$, and $\chi_{abs}$ are the absorption frequency, Planck constant, and local-field correction ($\chi_{abs} = \left\{(n^2+2)/3\right\}^2$ in the Lorentz model, where $n$ is refractive index). $\Omega_\lambda$ are the Judd-Ofelt parameters. The reduced matrix element $\left\langle l^N SLJ_{mixed}\middle\|U^{(\lambda)}\middle\|l^N S'L'J'_{mixed}\right\rangle$ can be expanded using Eq. (2.10) as:

$$\left\langle l^N SLJ_{mixed}\middle\|U^{(\lambda)}\middle\|l^N S'L'J'_{mixed}\right\rangle = \sum\nolimits_{i,j} c_i c_j \left\langle l^N SLJ_{pure}\middle\|U^{(\lambda)}\middle\|l^N S'L'J'_{pure}\right\rangle. \quad (2.14)$$

$\left\langle l^N SLJ_{pure}\middle\|U^{(\lambda)}\middle\|l^N S'L'J'_{pure}\right\rangle$ is expressed by tensor operator technique as:

$$\left\langle l^N SLJ\middle\|U^{(\lambda)}\middle\|l^N S'L'J'\right\rangle = (-1)^{S+L'+J+\lambda}\sqrt{(2J+1)(2J'+1)} \left\{\begin{matrix} J & J' & \lambda \\ L' & L & S \end{matrix}\right\}\left\langle l^N SL\middle\|U^{(\lambda)}\middle\|l^N SL'\right\rangle, \quad (2.15)$$

in which $\left\langle l^N SL\middle\|U^{(\lambda)}\middle\|l^N SL'\right\rangle$ can be further reduced to:

$$\left\langle l^N SL\middle\|U^{(\lambda)}\middle\|l^N SL'\right\rangle = N\sqrt{(2L+1)(2L'+1)} \sum\nolimits_{\psi(l^{N-1})}\left(l^{N-1}\overline{SL}|l^N SL\right)\left\{\begin{matrix} L & l & \bar{L} \\ l & L' & \lambda \end{matrix}\right\}(-1)^{\bar{L}+L+l+\lambda}. \quad (2.16)$$

Here, like Eq. (2.9):

$$\left\langle l^{14-N} SL\middle\|U^{(\lambda)}\middle\|l^{14-N} SL'\right\rangle = -(-1)^\lambda\left\langle l^N SL\middle\|U^{(\lambda)}\middle\|l^N SL'\right\rangle. \quad (2.17)$$

Meanwhile, the oscillator strength of magnetic dipole transition $f_{MD}$ is:

$$f_{MD} = \frac{hv_{abs}}{6m_e c^2}\frac{n}{2J+1}\left|\left\langle l^N SLJ_{mixed}\|L+gS\|l^N S'L'J'_{mixed}\right\rangle\right|^2. \quad (2.18)$$

Similarly to Eq. (2.13), the reduced matrix element can be expanded with pure wavefunctions. Magnetic-dipole transitions are allowed when $\Delta J = 0, \pm 1$. The reduced matrix element can be calculated by:

$$\left\langle l^N SLJ \| L + gS \| l^N SLJ' \right\rangle = (g-1)\sqrt{\frac{(S+L+J+1)(J+L-S)(J+S-L)(S+L-J+1)}{4J}}, \tag{2.19}$$

for $J' = J - 1$,

$$\left\langle l^N SLJ \| L + gS \| l^N SLJ' \right\rangle = g\sqrt{J(J+1)(2J+1)}, \tag{2.20}$$

for $J' = J$, and

$$\left\langle l^N SLJ \| L + gS \| l^N SLJ' \right\rangle = (g-1)\sqrt{\frac{(S+L+J+2)(J+L-S+1)(J+S-L+1)(S+L-J)}{4J}}, \tag{2.21}$$

for $J' = J + 1$, where g = 2.002319304362… is the $g$-factor of the electron [11]. Unlike the oscillator strength of electric dipole transition, there is no Judd-Ofelt parameter $\Omega_\lambda$. Therefore, the value of $f_{\mathrm{MD}}$ can be calculated theoretically unlike $f_{\mathrm{ED}}$, which requires optimization of Judd-Ofelt parameters $\Omega_\lambda$ from experimental oscillator strengths.

From the oscillator strength obtained from the absorption spectra of a lanthanide in a material under investigation, the value of total oscillator strength $f_{\mathrm{tot}}$ is derived:

$$f_{\mathrm{tot}} = f_{\mathrm{ED}} + f_{\mathrm{MD}}. \tag{2.22}$$

After obtaining oscillator strength for multiple transitions, they are fitted to Eq. (2.13) to yield an overdetermined system of equations. Through optimization, the values of the Judd-Ofelt parameters $\Omega_\lambda$ can be obtained.

The radiative rate constant $A_{\mathrm{rad}}$ is composed of electric dipole transition $A_{\mathrm{rad(ED)}}$ and magnetic dipole transition $A_{\mathrm{rad(MD)}}$ and they are given by the following equations:

$$A_{\mathrm{rad}} = A_{\mathrm{rad(ED)}} + A_{\mathrm{rad(MD)}}, \tag{2.23}$$

$$A_{\mathrm{rad(ED)}} = \frac{2\pi v_{\mathrm{em}}^2 e^2}{\varepsilon_0 c^3 m} n^2 f_{\mathrm{ED}}, \tag{2.24}$$

$$A_{\mathrm{rad(MD)}} = \frac{2\pi v_{\mathrm{em}}^2 e^2}{\varepsilon_0 c^3 m} n^2 f_{\mathrm{MD}}. \tag{2.25}$$

### 2.2.3 Energy Transfer Rate Constants

The derivation of energy transfer matrix element between two Ln(III) ions can be expressed by using the operator for Coulomb interaction between two Ln(III) ions. As stated in the beginning of this section, exchange interaction will be neglected. With tensor operator technique for diatomic molecules, the operator for resonance interaction is:

$$H_{\mathrm{AB}} = e^2 \sum_{\substack{k_1, k_2 \\ q_1, q_2}} (-1)^{k_1} \left\{ \frac{(2k_1 + 2k_2 + 1)}{(2k_1)!\,(2k_2)!} \right\}^{\frac{1}{2}} \begin{pmatrix} k_1 & k_2 & k_1 + k_2 \\ q_1 & q_2 & -(q_1 + q_2) \end{pmatrix} \frac{C_{q_1+q_2}^{k_1+k_2^*}}{R^{k_1+k_2+1}} D_{Aq_1}^{(k_1)} D_{Bq_2}^{(k_2)}, \tag{2.26}$$

where $R$ is the distance between atom A and B, and $D_{\mathrm{A}q_1}^{(k_1)}$ and $D_{\mathrm{B}q_2}^{(k_2)}$ are the electric multipole operator $\left(D_q^{(k)} = \sum_i r_i^k C_q^{(k)}\right)$ for A and B centers, respectively. Calculating the squared matrix elements using this operator leads to infinite series of interaction types, e.g., dipole-dipole, dipole-quadrupole (and vise-versa), quadrupole-quadrupole, quadrupole-hexadecapole, and so on [12]. However, the theoretical energy transfer rate constant reported by Kushida [3] under the formulation of Judd-Ofelt theory concluded that higher-order interactions above quadrupole-quadrupole could be neglected in most cases. Since the interaction between Ln(III) ions are very small, $H_{\mathrm{AB}}$ can be considered as a perturbation to the system. Inserting the matrix element of the energy transfer into the Fermi's Golden rule, the rate constant of the energy transfer is given by:

$$W_{\mathrm{ET}} = \frac{2\pi}{\hbar} |\langle AB|H_{\mathrm{AB}}|A'B'\rangle|^2 F, \tag{2.27}$$

where $F(= \int S_{A\rightarrow A'}(E) S_{B\rightarrow B'}(E) dE)$ is the spectral overlap between the transitions of A and B to A′ and B′, respectively. For dipole-dipole, dipole-quadrupole (and vice versa), and quadrupole-quadrupole interaction of energy transfer between two Ln(III) ions A and B, the rate constants are:

$$W_{\mathrm{ET(dd)}} = \frac{1}{(2J_{\mathrm{a}}+1)(2J_{\mathrm{b}}+1)} \frac{2}{3} \frac{4\pi^2}{h} \frac{e^4}{R^6} \left[ \sum_{\lambda=2,4,6} \Omega_\lambda^A \left|\langle l^N SLJ \| U^{(\lambda)} \| l^N S'L'J' \rangle\right|^2 \right] \times \left[ \sum_{\lambda=2,4,6} \Omega_\lambda^B \left|\langle l^N SLJ \| U^{(\lambda)} \| l^N S'L'J' \rangle\right|^2 \right] F, \tag{2.28}$$

$$W_{\mathrm{ET(dq)}} = \frac{1}{(2J_{\mathrm{a}}+1)(2J_{\mathrm{b}}+1)} \frac{4\pi^2}{h} \frac{e^4}{R^8} \left[ \sum_{\lambda=2,4,6} \Omega_\lambda^A \left| \langle l^N SLJ \| U^{(\lambda)} \| l^N S'L'J' \rangle \right|^2 \right]$$
$$\times \left[ \left| \langle 4\mathrm{f} \| r_B^2 \| 4\mathrm{f} \rangle \right|^2 \left| \langle l \| C^{(2)} \| l \rangle \right|^2 \left| \langle l^N SLJ \| U^{(2)} \| l^N S'L'J' \rangle \right|^2 \right] F, \tag{2.29}$$

$$W_{\mathrm{ET(qq)}} = \frac{1}{(2J_{\mathrm{a}}+1)(2J_{\mathrm{b}}+1)} \frac{14}{5} \frac{4\pi^2}{h} \frac{e^4}{R^{10}} \left| \langle l \| C^{(2)} \| l \rangle \right|^4$$
$$\left[ \left| \langle 4\mathrm{f} \| r_A^2 \| 4\mathrm{f} \rangle \right|^2 \left| \langle l^N SLJ \| U^{(2)} \| l^N S'L'J' \rangle \right|^2 \right]$$
$$\times \left[ \left| \langle 4\mathrm{f} \| r_B^2 \| 4\mathrm{f} \rangle \right|^2 \left| \langle l^N SLJ \| U^{(2)} \| l^N S'L'J' \rangle \right|^2 \right] F, \tag{2.30}$$

where $R$, $\langle l \| C^{(2)} \| l \rangle$, $\langle 4\mathrm{f} \| r_A^2 \| 4\mathrm{f} \rangle$, and $\langle l^N SLJ \| U^{(2)} \| l^N S'L'J' \rangle$ are the distance between Ln(III) ions, one-electron reduced matrix element, and reduced matrix elements of the tensor operator $U^{(2)}$ that is calculated from Eqs. (2.15) and (2.16). The mixed wavefunctions due to intermediate coupling must also be accounted for in the reduced matrix elements. $\langle l \| C^{(2)} \| l \rangle$ is calculated by the following equation:

$$\langle l \| C^{(2)} \| l \rangle = (-1)^l (2l+1) \begin{pmatrix} l & 2 & l \\ 0 & 0 & 0 \end{pmatrix}. \tag{2.31}$$

The values of $\langle 4\mathrm{f} \| r_A^2 \| 4\mathrm{f} \rangle$ for Ln(III) ions are provided in the literature [3]. $F$ is conveniently expressed as the overlap integral of emission and absorption given by the following equation [6]:

$$F = \frac{\ln 2}{\sqrt{\pi}} \frac{1}{\hbar\gamma_D \hbar\gamma_A} \sqrt{\left\{ \left(\frac{1}{\hbar\gamma_D}\right)^2 + \left(\frac{1}{\hbar\gamma_A}\right)^2 \right\} \ln 2} \times \exp\left[ \frac{1}{4} \frac{\left(\frac{2\Delta}{\hbar\gamma_D{}^2} \ln 2\right)^2}{\left\{ \left(\frac{1}{\hbar\gamma_D}\right)^2 + \left(\frac{1}{\hbar\gamma_A}\right)^2 \right\} \ln 2} - \left(\frac{\Delta}{\hbar\gamma_D}\right)^2 \ln 2 \right], \tag{2.32}$$

where $\hbar\gamma_{\mathrm{D}}$, $\hbar\gamma_{\mathrm{A}}$, and $\Delta$ are full-width at half-maximum (FWHM) of emission, FWHM of absorption, and the centroid-to-centroid gap between emission and absorption, respectively.

$W_{\mathrm{ET(dd)}}$ and $W_{\mathrm{ET(dq)}}$ (Eqs. (2.28) and (2.29)) contain the Judd-Ofelt parameter $\Omega_\lambda$. Under the assumption that the 4f-state wavefunctions are already optimized $|l^N SLJ_{\mathrm{mixed}}\rangle$), the energy transfer rate constants can still be calculated without $\Omega_\lambda$ because the equation only requires the value for the entire sum of the reduced matrix element. Looking at the total oscillator strength $f_{\mathrm{tot}}$ in Eq. (2.22), which can be derived experimentally through absorption spectroscopy, the oscillator strength of electric dipole transition $f_{\mathrm{ED}}$ can be determined once $f_{\mathrm{MD}}$ is calculated from Eq. (2.18). From Eq. (2.13), the sum of the matrix elements can be calculated, and thus the values for $W_{\mathrm{ET(dd)}}$ and $W_{\mathrm{ET(dq)}}$ as well.

## 2.3 Kinetic Analysis of Excited State

Kinetics dominate any photophysical processes in a material. In the simplest case (such as two-level system), it is sometimes easy to visualize the excited state dynamics. However, a process such as energy transfer that could cause interaction between emitting species to span over a large spatial area could lead to a complicated excited state dynamics. The photophysical properties of a system with interactions are sometimes hard to intuitively understand and analyze. This is especially common for an inhomogeneous system where local dynamics are significantly different from the other parts of the system. Kinetic analysis is a powerful tool to model and analyze a system mathematically [4].

Consider N species where each species $i$ (= 1, 2, ..., $N$) is transitioning from ground to excited state and vise-versa, with or without interaction with another species. Let all of the processes to be a linear processes, meaning that the order of process of all processes are one, in contrast to a nonlinear process where the order of process is not one. Quantum cutting is one example of a nonlinear process because it produces two excited states from one excited state in a single process. The population density of the entire system is constant and normalized in a closed system, that is:

$$\sum_i x_i(t) + \sum_i x_i'(t) = 1, \tag{2.33}$$

where $x_i(t)$ and $x_i'(t)$ are the ith species at excited and ground state, respectively. Let the population density of ground states be significantly larger than excited states (excitation source with reasonable power is being used), i.e. $\sum_i x_i(t) \ll \sum_i x_i'(t) = 1$ (steady state approximation). With these premises, a system of $N$ linear differential equations are expressed as:

$$\begin{aligned}
\tfrac{d}{dt}x_1(t) &= A_{1,1}x_1(t) + A_{2,1}x_2(t) + \cdots + A_{N,1}x_N(t),\\
\tfrac{d}{dt}x_2(t) &= A_{1,2}x_1(t) + A_{2,2}x_2(t) + \cdots + A_{N,2}x_N(t),\\
&\vdots\\
\tfrac{d}{dt}x_N(t) &= A_{1,N}x_1(t) + A_{2,N}x_2(t) + \cdots + A_{N,N}x_N(t),
\end{aligned} \tag{2.34}$$

where $A_{i,j}$ ($i \neq j$) is the interaction rate constant involving $x_i(t)$ and $x_j(t)$. $A_{i,i}$ is the relaxation rate constant of $x_i(t)$. At this point, the system is closed and there is no excitation source to produce any excited state. So we add input functions $J_i(t)$ to Eq. (2.34):

$$\begin{aligned}
\tfrac{d}{dt}x_1(t) &= A_{1,1}x_1(t) + A_{2,1}x_2(t) + \cdots + A_{N,1}x_N(t) + J_1(t),\\
\tfrac{d}{dt}x_2(t) &= A_{1,2}x_1(t) + A_{2,2}x_2(t) + \cdots + A_{N,2}x_N(t) + J_2(t)\\
&\vdots\\
\tfrac{d}{dt}x_N(t) &= A_{1,N}x_1(t) + A_{2,N}x_2(t) + \cdots + A_{N,N}x_N(t) + J_N(t).
\end{aligned} \tag{2.35}$$

This is now a system of inhomogeneous $N$ linear differential equations. It is possible to write Eq. (2.35) in the matrix form:

$$\frac{d}{dt}\boldsymbol{X}(t) = \boldsymbol{A}\boldsymbol{X}(t) + \boldsymbol{J}(t), \tag{2.36}$$

where

$$\boldsymbol{X}(t) = \begin{pmatrix} x_1(t) \\ x_2(t) \\ \vdots \\ x_N(t) \end{pmatrix}, \tag{2.37}$$

$$\boldsymbol{A} = \begin{pmatrix} A_{1,1} & A_{2,1} & \cdots & A_{N,1} \\ A_{1,2} & A_{2,2} & & \vdots \\ \vdots & & \ddots & \vdots \\ A_{1,N} & \cdots & \cdots & A_{N,N} \end{pmatrix}, \tag{2.38}$$

$$\boldsymbol{J}(t) = \begin{pmatrix} J_1(t) \\ J_2(t) \\ \vdots \\ J_N(t) \end{pmatrix}. \tag{2.39}$$

There is no mathematical difficulty in deriving Eq. (2.36). The matrix multiplication in Eq. (2.36) will easily yield Eq. (2.35). Let the excitation begin at $t = 0$, and the analytical solution of (2.36) is:

$$\boldsymbol{X}(t) = \int_0^t \exp\{(t - t')\boldsymbol{A}\}\boldsymbol{J}(t')\mathrm{d}t'. \tag{2.40}$$

Using this equation, it is possible to solve for $\boldsymbol{X}(t)$ with $\boldsymbol{J}(t)$ of impulse function, corresponding to a pulse excitation, or a constant, corresponding to a steady state excitation.

### 2.3.1 Rate Equation for Impulse Excitation

Solving the rate equation $\boldsymbol{X}(t)$ with $\boldsymbol{J}(t)$ of impulse function leads to population density directly related to emission lifetimes. Emission lifetimes are linearly proportional to the population density. Let only the species $i = 1$ is excited, and then the input function $\boldsymbol{J}(t)$ can be expressed as:

$$\boldsymbol{J}(t) = \begin{pmatrix} J_1(t) \\ J_2(t) \\ \vdots \\ J_N(t) \end{pmatrix} = j_1 \begin{pmatrix} \delta(t) \\ 0 \\ \vdots \\ 0 \end{pmatrix}, \tag{2.41}$$

where $j_1$ is the amplitude of the input, which for convenience will be set to 1. The impulse function, or the Dirac delta function $\delta(x)$ is defined (actually one of the definitions) as:

$$\int_{a-\epsilon}^{a+\epsilon} f(x)\delta(x-a)dx = f(a) \quad (\epsilon > 0). \tag{2.42}$$

Using Eq. (2.41), Eq. (2.39) with input function defined in Eq. (2.40) is further solved to give:

$$\boldsymbol{X}(t) = \exp(\boldsymbol{A}t)\boldsymbol{B}, \tag{2.43}$$

where

$$\boldsymbol{B} = \begin{pmatrix} j_1 \\ 0 \\ \vdots \\ 0 \end{pmatrix} = \begin{pmatrix} 1 \\ 0 \\ \vdots \\ 0 \end{pmatrix}. \tag{2.44}$$

$\exp(\boldsymbol{A}t)$ is called matrix exponential. Equation (2.43) is now an eigenvalue problem and diagonalization of the matrix exponential leads to the rate equations for each $x_i(t)$. Equation (2.42) may not be solved analytically for $N > 4$ with some exception. However, unlike the numerical methods to solve for differential equation in the form of Eq. (2.35) (such as 4th-order Runge-Kutta method), many numerical (iterative) methods for the diagonalization of matrices are comparatively accurate and faster.

### 2.3.2 Rate Equation for Constant Excitation

Solving the rate equation $\boldsymbol{X}(t)$ with $\boldsymbol{J}(t)$ of constant leads to population density related to quantum yields. Let only the species $i = 1$ is excited, and then the input function $\boldsymbol{J}(t)$ can be expressed as:

$$\boldsymbol{J}(t) = \begin{pmatrix} J_1(t) \\ J_2(t) \\ \vdots \\ J_N(t) \end{pmatrix} = j_1 \begin{pmatrix} 1 \\ 0 \\ \vdots \\ 0 \end{pmatrix}, \tag{2.45}$$

where $j_1$ is the amplitude of the input, which for convenience will be set to 1 once again. Equation (2.40) with input function defined in Eq. (2.45) is further solved to give:

$$\boldsymbol{X}(t) = \boldsymbol{A}^{-1}\{\exp(\boldsymbol{A}t) - \boldsymbol{I}\}\boldsymbol{B}, \tag{2.46}$$

where

$$\boldsymbol{B} = \begin{pmatrix} j_1 \\ 0 \\ \vdots \\ 0 \end{pmatrix} = \begin{pmatrix} 1 \\ 0 \\ \vdots \\ 0 \end{pmatrix}. \tag{2.47}$$

### 2.3.3 Simple Mononuclear Case

The simplest form of mononuclear Ln(III) complex is shown in Fig. 2.2a and the simplest mechanism is shown in Fig. 2.2b as a 4-level system (same as Fig. 1.7a, c of Chap. 1). The rate equations are derived for this simple case with BET as an example of kinetic analysis. Absorption of a photon leads to a transition from singlet ground state $S_0$ of the ligand to singlet excited state $S_1$ and is immediately followed by intersystem crossing to the triplet excited state $T_1$. Energy transfer from the $T_1$ state to two-level Ln(III) ion leads to excited state of Ln(III) ion (Ln*). The rate equation for this system is:

$$\frac{d}{dt}\boldsymbol{X}(t) = \boldsymbol{A}\boldsymbol{X}(t) + \boldsymbol{J}(t), \tag{2.36}$$

$$\boldsymbol{X}(t) = \begin{pmatrix} \mathrm{S}_1(t) \\ \mathrm{T}_1(t) \\ \mathrm{Ln}^*(t) \end{pmatrix}, \tag{2.48}$$

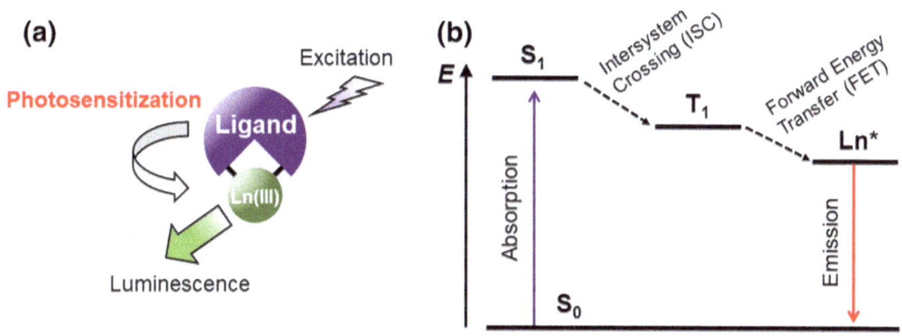


**Fig. 2.2** **a** Schematic depiction of Ln(III) complexes. **b** Luminescence mechanism of Ln(III) complexes. Ln* means an excited Ln(III) ion

$$\boldsymbol{A} = \begin{pmatrix} -a_{11} & 0 & 0 \\ a_{21} & -a_{22} & a_{23} \\ 0 & a_{32} & -a_{33} \end{pmatrix}, \tag{2.49}$$

where

$$a_{11} = kr_{\mathrm{S_1}} + knr_{\mathrm{S_1}} + k_{\mathrm{ISC}}, \tag{2.50}$$

$$a_{21} = k_{\mathrm{ISC}}, \tag{2.51}$$

$$a_{22} = kr_{\mathrm{T_1}} + knr_{\mathrm{T_1}} + k_{\mathrm{FET}}, \tag{2.52}$$

$$a_{23} = k_{\mathrm{BET}}, \tag{2.53}$$

$$a_{33} = kr_{\mathrm{Ln^*}} + knr_{\mathrm{Ln^*}} + k_{\mathrm{BET}}, \tag{2.54}$$

$$a_{32} = k_{\mathrm{FET}}, \tag{2.55}$$

and $k_{\mathrm{ISC}}$, $kr$, $knr$, $k_{\mathrm{FET}}$, and $k_{\mathrm{BET}}$ are defined as intersystem-crossing, radiative, non-radiative, FET, and BET, respectively. The subscript indicates a species.

For both the impulse and constant excitations, analytical solution is available because solving for $\boldsymbol{X}(t)$ is an eigenvalue problem of $3 \times 3$ square matrix. Eigenvalue problem of $3 \times 3$ square matrix is essentially about solving the cubic equation in which an algebraic solution is available. For impulse excitation, the rate equations are:

$$S_1(t) = \exp(-a_{11}t), \tag{2.56}$$

$$\begin{aligned}
T_1(t) = {} & \frac{a_{21}}{4\left(a_{11}a_{22} + a_{11}a_{33} - a_{22}a_{33} + a_{23}a_{32} - a_{11}^2\right)} \{4(a_{11} - a_{33})\exp(-a_{11}t) \\
& + \frac{\left(-a_{22} + a_{33} - \sqrt{a_{22}^2 - 2a_{22}a_{33} + a_{33}^2 + 4a_{23}a_{32}}\right)\left(2a_{11} - a_{22} - a_{33} + \sqrt{a_{22}^2 - 2a_{22}a_{33} + a_{33}^2 + 4a_{23}a_{32}}\right)}{\sqrt{a_{22}^2 - 2a_{22}a_{33} + a_{33}^2 + 4a_{23}a_{32}}} \\
& \exp\left(-\frac{a_{22} + a_{33} + \sqrt{a_{22}^2 - 2a_{22}a_{33} + a_{33}^2 + 4a_{23}a_{32}}}{2}t\right) \\
& - \frac{\left(a_{22} + a_{33} + \sqrt{a_{22}^2 - 2a_{22}a_{33} + a_{33}^2 + 4a_{23}a_{32}}\right)\left(-2a_{11} + a_{22} + a_{33} + \sqrt{a_{22}^2 - 2a_{22}a_{33} + a_{33}^2 + 4a_{23}a_{32}}\right)}{\sqrt{a_{22}^2 - 2a_{22}a_{33} + a_{33}^2 + 4a_{23}a_{32}}} \\
& \exp\left(-\frac{a_{22} + a_{33} - \sqrt{a_{22}^2 - 2a_{22}a_{33} + a_{33}^2 + 4a_{23}a_{32}}}{2}t\right)\Bigg\}.
\end{aligned} \tag{2.57}$$

$$\begin{aligned}
\mathrm{Ln}^*(t) = {} & \frac{a_{21}a_{32}}{2\left(a_{11}a_{22} + a_{11}a_{33} - a_{22}a_{33} + a_{23}a_{32} - a_{11}^2\right)} \\
& \Bigg\{-2\exp(-a_{11}t) + \frac{2a_{11} - a_{22} - a_{33} + \sqrt{a_{22}^2 - 2a_{22}a_{33} + a_{33}^2 + 4a_{23}a_{32}}}{\sqrt{a_{22}^2 - 2a_{22}a_{33} + a_{33}^2 + 4a_{23}a_{32}}} \\
& \exp\left(-\frac{a_{22} + a_{33} + \sqrt{a_{22}^2 - 2a_{22}a_{33} + a_{33}^2 + 4a_{23}a_{32}}}{2}t\right) \\
& + \frac{-2a_{11} + a_{22} + a_{33} + \sqrt{a_{22}^2 - 2a_{22}a_{33} + a_{33}^2 + 4a_{23}a_{32}}}{\sqrt{a_{22}^2 - 2a_{22}a_{33} + a_{33}^2 + 4a_{23}a_{32}}} \\
& \exp\left(-\frac{a_{22} + a_{33} - \sqrt{a_{22}^2 - 2a_{22}a_{33} + a_{33}^2 + 4a_{23}a_{32}}}{2}t\right)\Bigg\}.
\end{aligned} \tag{2.58}$$

These population density equations are exponential equations. The emission decay profile and population density are equivalent regarding lifetimes since emission intensity is proportional to the population density. Thus the emission lifetime is defined as the amount of time it takes for the intensity to drop to $1/e$ of the initial intensity. The equations are quite complicated but by breaking down the equations into terms and interpreting them each, the equations provide meaningful information. Equation (2.56) is a straight forward equation that represents the time evolution of the $S_1$ state. Equation (2.57) has three terms. The first term represents the contribution of $S_1$ state decay and corresponds to the rise of the time evolution of the $T_1$ state.

The second term represents the decay of the $T_1$ state. The last term represents the contribution of the BET from the Ln* state. For Eq. (2.58), it can be seen that the BET contributes directly into the Ln* state, similar to the second term of Eq. (2.57). BET mixes pure $T_1$ and Ln* state dynamics into mixed dynamics. In the absence of BET, the equations become much simpler:

$$S_1(t) = \exp(-a_{11}t), \tag{2.56}$$

$$T_1(t) = \frac{a_{21}}{2(a_{22} - a_{33})\left(a_{11}a_{22} + a_{11}a_{33} - a_{22}a_{33} - a_{11}^2\right)} \{2(a_{22} - a_{33})(a_{11} - a_{33})\exp(-a_{11}t) -2(a_{11} - a_{33})\exp(-a_{22}t) + a_{22}(a_{11} - a_{22})\exp(-a_{33}t)\}, \tag{2.57$'$}$$

$$\mathrm{Ln}^*(t) = \frac{a_{21}a_{32}}{(a_{22} - a_{33})\left(a_{11}a_{22} + a_{11}a_{33} - a_{22}a_{33} - a_{11}^2\right)} \{-(a_{22} - a_{33})\exp(-a_{11}t) + (a_{11} - a_{33})\exp(-a_{22}t) -(a_{11} - a_{22})\exp(-a_{33}t)\} \tag{2.58$'$}$$

Now the exponential terms of these three equations are pure. In most cases, $S_1$ and $T_1$ states have very short lifetimes ($a_{11} \gg a_{22} \gg a_{33}$), and Eq. (2.58′) can be further simplified to:

$$\mathrm{Ln}^*(t) = \frac{a_{21}a_{32}}{(a_{22} - a_{33})(a_{11} - a_{33})}\exp(-a_{33}t) = \mathrm{Ln}^*(0)\exp(-a_{33}t). \tag{2.58$''$}$$

This equation is now the commonly seen emission decay with lifetime of $\frac{1}{a_{33}} = \frac{1}{kr_{\mathrm{Ln}^*} + knr_{\mathrm{Ln}^*}} = \tau_{\mathrm{obs}}$ in literature. Equation (2.58″) is merely an equation derived with the assumption that BET does not occur and the $S_1$ and $T_1$ states have short lifetimes compared to the excited state of Ln(III) ion. Sometimes, however, $T_1$ states may have a long lifetime (still shorter than Ln(III)) with those of Ln(III) ion. In that case, the emission decay profile of Ln(III) will have an initial rise in the intensity corresponding to the donor lifetime (=$T_1$ state).

For constant excitation, the rate equations are:

$$S_1(t) = -\frac{1}{a_{11}}\{\exp(-a_{11}t) - 1\}, \tag{2.59}$$

$$
\begin{aligned}
\mathrm{T}_1(t) = -\frac{a_{21}}{a_{22}a_{33}-a_{23}a_{32}} \Bigg[ & \frac{a_{33}}{4\left(a_{11}a_{22}-a_{11}^2+a_{11}a_{33}-a_{22}a_{33}+a_{23}a_{32}\right)} \\
& \{4(a_{11}-a_{33})\exp(-a_{11}t) \\
& -\frac{\left(a_{22}-a_{33}+\sqrt{a_{22}^2-2a_{22}a_{33}+a_{33}^2+4a_{23}a_{32}}\right)\left(2a_{11}-a_{22}-a_{33}+\sqrt{a_{22}^2-2a_{22}a_{33}+a_{33}^2+4a_{23}a_{32}}\right)}{\sqrt{a_{22}^2-2a_{22}a_{33}+a_{33}^2+4a_{23}a_{32}}} \\
& \exp\left(-\frac{a_{22}+a_{33}-\sqrt{a_{22}^2-2a_{22}a_{33}+a_{33}^2+4a_{23}a_{32}}}{2}t\right) \\
& -\frac{\left(a_{22}-a_{33}-\sqrt{a_{22}^2-2a_{22}a_{33}+a_{33}^2+4a_{23}a_{32}}\right)\left(a_{22}-2a_{11}+a_{33}+\sqrt{a_{22}^2-2a_{22}a_{33}+a_{33}^2+4a_{23}a_{32}}\right)}{\sqrt{a_{22}^2-2a_{22}a_{33}+a_{33}^2+4a_{23}a_{32}}} \\
& \exp\left(-\frac{a_{22}+a_{33}-\sqrt{a_{22}^2-2a_{22}a_{33}+a_{33}^2+4a_{23}a_{32}}}{2}t\right)\Bigg\} \\
& +\frac{a_{23}a_{32}}{2\left(a_{11}a_{22}-a_{11}^2+a_{11}a_{33}-a_{22}a_{33}+a_{23}a_{32}\right)} \\
& \left\{-2\exp(-a_{11}t)\frac{2a_{11}-a_{22}-a_{33}+\sqrt{a_{22}^2-2a_{22}a_{33}+a_{33}^2+4a_{23}a_{32}}}{\sqrt{a_{22}^2-2a_{22}a_{33}+a_{33}^2+4a_{23}a_{32}}}\right. \\
& \exp\left(-\frac{a_{22}+a_{33}+\sqrt{a_{22}^2-2a_{22}a_{33}+a_{33}^2+4a_{23}a_{32}}}{2}t\right) \\
& +\frac{a_{22}-2a_{11}+a_{33}+\sqrt{a_{22}^2-2a_{22}a_{33}+a_{33}^2+4a_{23}a_{32}}}{\sqrt{a_{22}^2-2a_{22}a_{33}+a_{33}^2+4a_{23}a_{32}}} \\
& \left.\exp\left(-\frac{a_{22}+a_{33}-\sqrt{a_{22}^2-2a_{22}a_{33}+a_{33}^2+4a_{23}a_{32}}}{2}t\right)\right\} + \frac{a_{33}}{a_{11}}\{\exp(-a_{11}t)-1\}\Bigg],
\end{aligned}
\tag{2.60}
$$

$$
\begin{aligned}
\mathrm{Ln}^*(t) = -\frac{a_{21}a_{32}}{a_{22}a_{33}-a_{23}a_{32}} \Bigg[ & \frac{1}{4\left(a_{11}a_{22}+a_{11}a_{33}-a_{22}a_{33}+a_{23}a_{32}-a_{11}^2\right)} \\
& \{4(-a_{11}t)(a_{11}-a_{33})\exp(-a_{11}t) \\
& -\frac{\left(a_{22}-a_{33}+\sqrt{a_{22}^2-2a_{22}a_{33}+a_{33}^2+4a_{23}a_{32}}\right)\left(2a_{11}-a_{22}-a_{33}+\sqrt{a_{22}^2-2a_{22}a_{33}+a_{33}^2+4a_{23}a_{32}}\right)}{\sqrt{a_{22}^2-2a_{22}a_{33}+a_{33}^2+4a_{23}a_{32}}}\Bigg\} \\
& \exp\left(-\frac{a_{22}+a_{33}+\sqrt{a_{22}^2-2a_{22}a_{33}+a_{33}^2+4a_{23}a_{32}}}{2}t\right) \\
& -\frac{\left(a_{22}-a_{33}-\sqrt{a_{22}^2-2a_{22}a_{33}+a_{33}^2+4a_{23}a_{32}}\right)\left(a_{22}-2a_{11}+a_{33}+\sqrt{a_{22}^2-2a_{22}a_{33}+a_{33}^2+4a_{23}a_{32}}\right)}{\sqrt{a_{22}^2-2a_{22}a_{33}+a_{33}^2+4a_{23}a_{32}}} \\
& \exp\left(-\frac{a_{22}+a_{33}-\sqrt{a_{22}^2-2a_{22}a_{33}+a_{33}^2+4a_{23}a_{32}}}{2}t\right) \\
& +\frac{a_{22}}{2\left(a_{11}a_{22}-a_{11}^2+a_{11}a_{33}-a_{22}a_{33}+a_{23}a_{32}\right)} \\
& \{-2\exp(-a_{11}t)
\end{aligned}
$$

$$+\frac{1}{2\left(a_{11}-a_{22}-a_{33}+\sqrt{a_{22}^2-2a_{22}a_{33}+a_{33}^2+4a_{23}a_{32}}\right)+\sqrt{a_{22}^2-2a_{22}a_{33}+a_{33}^2+4a_{23}a_{32}}}\Bigg\}$$

$$\exp\left(-\frac{a_{22}+a_{33}+\sqrt{a_{22}^2-2a_{22}a_{33}+a_{33}^2+4a_{23}a_{32}}}{2}t\right)$$

$$+\frac{1}{\left(a_{22}-2a_{11}+a_{33}+\sqrt{a_{22}^2-2a_{22}a_{33}+a_{33}^2+4a_{23}a_{32}}\right)\sqrt{a_{22}^2-2a_{22}a_{33}+a_{33}^2+4a_{23}a_{32}}}$$

$$\exp\left(-\frac{a_{22}+a_{33}-\sqrt{a_{22}^2-2a_{22}a_{33}+a_{33}^2+4a_{23}a_{32}}}{2}t\right)\Bigg\}+\frac{1}{a_{11}}\{\exp(-a_{11}t)-1\}\Bigg]. \tag{2.61}$$

The rate equation seems complicated at first, but considering that the emission lifetime of Ln(III) ions is in the order of milliseconds, at sufficient elapsed time (order of seconds), above equations reduce to:

$$S_1=\frac{1}{a_{11}}=\frac{1}{kr_{S_1}+knr_{S_1}+k_{ISC}}, \tag{2.59'}$$

$$\begin{aligned}T_1&=\frac{a_{21}a_{33}}{a_{11}(a_{22}a_{33}-a_{23}a_{32})}=\frac{k_{ISC}}{kr_{S_1}+knr_{S_1}+k_{ISC}}\\&\times\frac{kr_{Ln^*}+knr_{Ln^*}+k_{BET}}{(kr_{T_1}+knr_{T_1}+k_{FET})(kr_{Ln^*}+knr_{Ln^*}+k_{BET})-k_{BET}k_{FET}}=\eta_{ISC}\\&\times\frac{kr_{Ln^*}+knr_{Ln^*}+k_{BET}}{(kr_{T_1}+knr_{T_1}+k_{FET})(kr_{Ln^*}+knr_{Ln^*}+k_{BET})-k_{BET}k_{FET}},\end{aligned} \tag{2.60'}$$

$$\begin{aligned}Ln^*&=\frac{a_{21}a_{32}}{a_{11}(a_{22}a_{33}-a_{23}a_{32})}=\frac{k_{ISC}}{kr_{S_1}+knr_{S_1}+k_{ISC}}\\&\times\frac{k_{FET}}{(kr_{T_1}+knr_{T_1}+k_{FET})(kr_{Ln^*}+knr_{Ln^*}+k_{BET})-k_{BET}k_{FET}}=\eta_{ISC}\\&\times\frac{k_{FET}}{(kr_{T_1}+knr_{T_1}+k_{FET})(kr_{Ln^*}+knr_{Ln^*}+k_{BET})-k_{BET}k_{FET}}.\end{aligned} \tag{2.61'}$$

The obtained rate equations (now constant at all time) each represents the (average) amount of time the species requires until any photophysical process proceeds. Therefore, by multiplying the rate constant of the particular process in question to the rate equation, the efficiency of the process can be derived. In the absence of BET, the equations above are simplified to:

$$\mathrm{S}_1 = \frac{1}{kr_{\mathrm{S}_1} + knr_{\mathrm{S}_1} + k_{\mathrm{ISC}}}, \tag{2.59''}$$

$$\mathrm{T}_1 = \eta_{\mathrm{ISC}} \times \frac{1}{\left(kr_{\mathrm{T}_1} + knr_{\mathrm{T}_1} + k_{\mathrm{FET}}\right)}, \tag{2.60''}$$

$$\begin{aligned} \mathrm{Ln}^* &= \eta_{\mathrm{sens}} \times \frac{k_{\mathrm{FET}}}{\left(kr_{\mathrm{T}_1} + knr_{\mathrm{T}_1} + k_{\mathrm{FET}}\right)\left(kr_{\mathrm{Ln}^*} + knr_{\mathrm{Ln}^*}\right)} \\ &= \eta_{\mathrm{ISC}} \times \eta_{\mathrm{FET}} \times \frac{1}{kr_{\mathrm{Ln}^*} + knr_{\mathrm{Ln}^*}} = \eta_{\mathrm{sens}} \times \frac{1}{kr_{\mathrm{Ln}^*} + knr_{\mathrm{Ln}^*}}. \end{aligned} \tag{2.61''}$$

By multiplying the radiative rate constant of Ln(III) ion $kr_{\mathrm{Ln}^*}$, Eq. (2.61″) becomes:

$$\mathrm{Ln}^* kr_{\mathrm{Ln}^*} = \eta_{\mathrm{sens}} \times \frac{kr_{\mathrm{Ln}^*}}{kr_{\mathrm{Ln}^*} + knr_{\mathrm{Ln}^*}} = \eta_{\mathrm{sens}} \times \Phi_{\mathrm{Ln}} = \Phi_{\mathrm{ligand}}. \tag{2.62}$$

Now the equation is equivalent to Eq. (1.3) in Chap. 1 and defines the quantum yield of the Ln(III) complexes.

As can be seen from these derivations, the equations for emission lifetime and quantum yield are based on many assumptions that is often not mentioned, deliberately omitted, or simply forgotten in most the literate. The commonly used equations can be easily invalidated in the presence of BET, and if they were to be used nonetheless, then there needs to be a legitimate basis to assume the validity in the particular model in question. For complicated models such as those described in this thesis, these kinetic analyses become necessary since the rate equations become sophisticated and the oversimplified model shown in this section may or may not apply.

## References

1. Judd BR (1962) Optical absorption intensities of rare-earth ions. Phys Rev 127:750–761
2. Ofelt GS (1962) Intensities of crystal spectra of rare-earth ions. J Chem Phys 37:511–520
3. Kushida T (1973) Energy transfer and cooperative optical transitions in rare-earth doped inorganic materials. I. transition probability calculation. J Phys Soc Jpn 34:1327–1333
4. Starzak ME (1989) Mathematical methods in chemistry and physics. Plenum Press, New York and London
5. Hehlen MP, Brik MG, Krämer KW (2013) 50th anniversary of the Judd-Ofelt theory: an experimentalist's view of the formalism and its application. J Lumin 136:221–239
6. Malta OL (2008) Mechanisms of non-radiative energy transfer involving lanthanide ions revisited. J Non Cryst Solids 354:4770–4776
7. Judd BR (2014) Operator techniques in atomic spectroscopy. Princeton University Press, New Jersey
8. Nielson CW, Koster GF (1963) Spectroscopic coefficients for the $p^n$, $d^n$, and $f^n$ configurations. M.I.T. Press
9. Sakurai JJ, Napolitano J (2011) Modern quantum mechanics 2nd edn. Pearson

10. Dieke GH, Crosswhite HM, Dunn B (1961) The spectra of the doubly and triply ionized rare earths. Appl Opt 2:675–686
11. Weber MJ (1967) Probabilities for Radiative and Nonradiative Decay of $Er^{3+}$ in $LaF_3$. Phys Rev 157:262–272
12. Dibartolo B (2012) Energy transfer processes in condensed matter. Plenum Press, New York

# Chapter 3
# Concentration Quenching in Ytterbium Coordination Polymers

**Abstract** Concentration quenching due to energy transfer between lanthanide ions have long been the shortcoming of lanthanide-based compounds, especially for luminescent functionalities like upconversion and multi-emission where energy transfer is necessary. Concentration quenching in lanthanide coordination polymers, on the other hand, is empirically known to be a significantly smaller effect than the inorganic counterparts. Elucidating the mechanism of concentration quenching in coordination polymers is therefore crucial for development of efficient functional luminescent materials. This chapter describes the concentration quenching mechanism of $[Yb(hfa)_3dpbp]_n$ (hfa: hexafluoroacetylacetonate, dpbp: 4,4′-bis(diphenylphosphoryl)-biphenyl) by experimental and theoretical approach. The mechanism of concentration quenching is found to be dominated by relaxation through phonon-assisted energy transfer.

## 3.1 Introduction

In Sect. 1.1, the energy transfer (ET) between Ln(III) ions and the unique properties arising from it was briefly mentioned. However, one of the major consequences of ET between Ln(III) ions is the concentration quenching [1, 2]. Concentration quenching has been extensively studied in Ln(III)-doped ceramics, and are known to reduce the emission quantum yield significantly [3–7]. For example, L. G. Jacobsohn showed that bulk $Tb:Y_2O_3$ with 1% Tb(III) concentration exhibit only half of the emission intensity of that of 0.5% Tb(III) concentration [8]. At 5% Tb(III) concentration, the intensity falls further down to 5%. Most commercially available Ln(III) doped materials thus contain only a few percent of Ln(III) ions to prevent the ET between them. Meanwhile, Ln(III) complexes, specifically Ln(III) coordination polymers, are known to show the less pronounced effect of concentration quenching [9–11]. Elucidation of the concentration quenching mechanism in Ln(III) coordination polymers could become important not only in achieving insights into the fundamental differences between Ln(III) coordination polymers and Ln(III)-doped ceramics, but also in obtaining strategies to suppress concentration quenching.



S. Omagari, *Energy Transfer Processes in Polynuclear Lanthanide Complexes*, Springer Theses, https://doi.org/10.1007/978-981-13-6049-7_3

There are mainly four models for concentration quenching: self-trapping model, killer-site model, cross-relaxation model, and phonon-assisted ET relaxation (PAETR) as depicted in Fig. 3.1 [3, 12]. In the self-trapping model (Fig. 3.1a), the emitted light of a Ln(III) ion is reabsorbed by the same species [3]. The killer-site model is characterized by nonradiative ET between multiple Ln(III) ions (energy migration) to "killer-sites" (impurities and defect sites, Fig. 3.1b [12, 13]. The self-quenching model can be further divided into limited-diffusion and fast-diffusion cases. The former refers to the case in which energy migration rate is comparable to the decay rate of the Ln(III) ions and is locally quenched. The latter refers to the case in which energy migration rate is much faster than the decay rate of the Ln(III) ions and the excited state spans over a large spatial area, probabilistically. Cross-relaxation occurs when ET from a Ln(III) ion with multiple energy levels populates different levels of identical Ln(III) ion and non-radiatively decays to the ground state (Fig. 3.1c) [14]. Finally, PAETR is a process where the energy from a donor is directly transferred to the phonon modes of the acceptor (Fig. 3.1d) [15]. This was proposed by F. Auzel in 2002 under the framework of phonon-assisted ET by Miyakawa and Dexter [16]. Concentration quenching in a material is usually comprised of more than one of these mechanisms and must be separately considered.

Experimental determination of the mechanism of concentration quenching in a material usually relies on Ln(III) concentration dependence of emission lifetimes instead of emission quantum yield [3–7]. Quantum yield only shows the number of emitted photons per absorbed photons, and thus it will simply decrease in value with increasing concentration of Ln(III) ions regardless of the mechanism. There is also an intrinsic problem in this method when used for Ln(III) coordination polymers because the ligands are excited instead of Ln(III) ions. This means that a change in Ln(III) concentration inevitably changes ligand-to-Ln(III) sensitization efficiency and thus

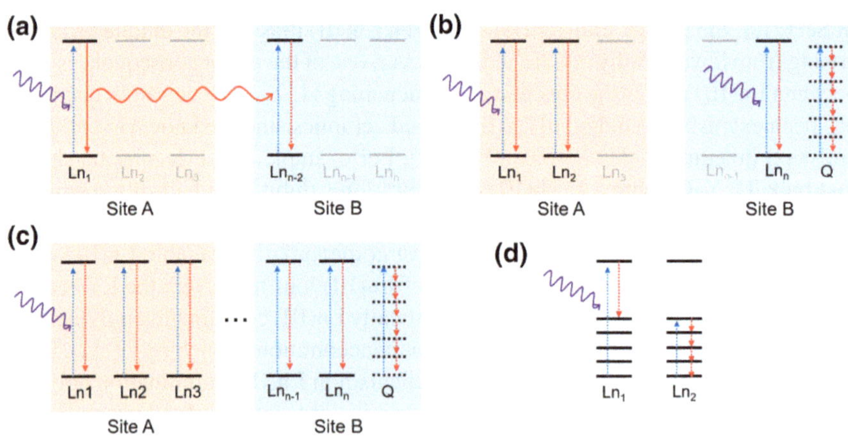


**Fig. 3.1** Concentration quenching models reported for Ln(III)-doped ceramics. **a** Self-trapping, **b** killer-site (KS), **c** cross-relaxation, and **d** phonon-assisted ET relaxation (PAETR). Reprinted from ref [44] with permission from John Wiley & Sons LTD

the quantum yield. On the other hand, emission lifetimes (or more precisely, emission decay profiles) provide information on the excited state dynamics as long as the dynamics involve the Ln(III) ions that is being monitored. For example, self-trapping is known to increase the emission lifetimes because there is a time lag between emitting and absorbing Ln(III) ions without any lifetime changes in the individual Ln(III) ions. In the limited-diffusion scheme of the killer-site model, the lifetimes of Ln(III) ions surrounding the killer-site is decreased while other Ln(III) far from the killer-site remains the same, leading to a multi-exponential decay profile [17]. In this study, relying mainly on emission lifetimes, the mechanism of concentration quenching in Ln(III) coordination polymers is elucidated.

Some reports consider concentration quenching in Ln(III) coordination polymers to arise from cross-quenching [10]. This is indeed a plausible explanation when considering Ln(III) ions with multiple excited states such as Er(III) ions. However, a report by V. V. Utochnikova and her coworkers indicate that concentration quenching occurs for coordination polymers with Yb(III) ions, which only have two 4f-states, namely the ground state $^2F_{7/2}$ and the excited state $^2F_{7/2}$ [18]. This two-level system simplifies ET from the ligands, ET between Yb(III) ions (YbYbET), and emission from the Yb(III) ions, and thus cross-quenching can be omitted from consideration. Thus, cross-quenching is not necessarily the only reason for concentration quenching. Furthermore, the nonradiative rate constant is generally large in Ln(III) coordination polymers, and this results in the lowered contribution of concentration quenching as well. Thus a small decrease in the emission lifetimes with increasing concentration of Ln(III) ions could be an indirect effect caused by the large nonradiative rate constant and not the fundamental property of Ln(III) coordination polymers. Judging concentration quenching as a small effect in Ln(III) coordination polymers based on a small decrease in the emission lifetimes could be a poor practice. This chapter focuses on $[Ln(hfa)_3dpbp]_n$ (hfa: hexafluoroacetylacetonate, dpbp: 4,4′-bis(diphenylphosphoryl)-biphenyl) to investigate Yb(III) concentration quenching [19, 20]. First, the energy transfer between Ln(III) ions (LnLnET) has been confirmed to occur for this system, which means that the concentration quenching is also likely to be observed. Second, the minimum distance between Ln(III) ions is 10.4 Å, allowing complete omission of any contribution from exchange interactions.

We synthesized $[Yb_xGd_{1-x}(hfa)_3dpbp]_n$ ($x = 0.001, 0.005, 0.01, 0.05, 0.1, 0.5, 1$, product of Fig. 3.2) and investigated the effect of concentration quenching of Yb(III) emission. The periodic structure and the local coordination geometry around Yb(III) ions of the prepared coordination polymers were characterized by powder XRD and diffuse reflectance spectra of the 4f–4f transition of Yb(III) ions, respectively. Emission spectra and lifetimes were measured to elucidate the concentration quenching mechanisms. It is suggested that the concentration mechanism in $[Yb_xGd_{1-x}(hfa)_3dpbp]_n$ is the self-generated quenching center by the PAETR, which was suggested in the report of F. Auzel in 2002. This is further qualitatively supported from theoretical calculations of the emission lifetimes. This investigation shows the fundamental difference between concentration quenching in Ln(III)-doped inorganic hosts and Ln(III) coordination polymers, and also provides a strategy to reduce the concentration quenching in Ln(III) coordination polymers.

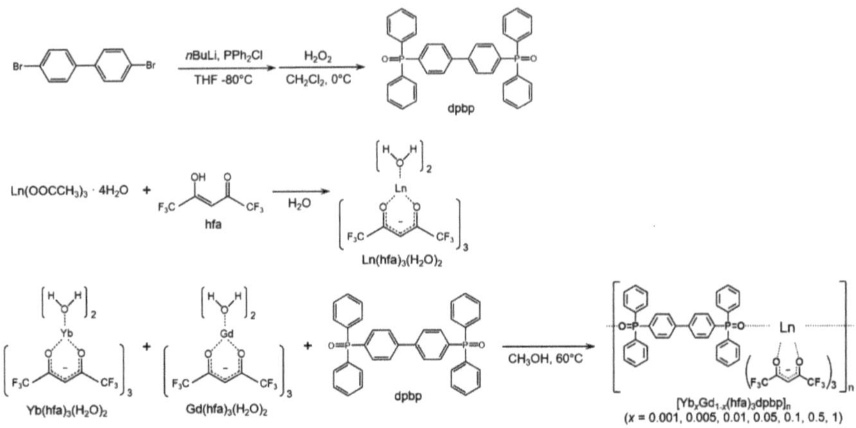


**Fig. 3.2** General synthetic method of $[Yb_xGd_{1-x}(hfa)_3dpbp]_n$ coordination polymers (three step process). Ln = Yb or Gd

## 3.2 Experimental Section

### *3.2.1 Material*

$Yb(OOCCH_3)_3 \cdot 4H_2O$ (99.9%) were purchased from Sigma-Aldrich Co. $Gd(OOCCH_3)_3 \cdot 4H_2O$ (>99.9) were purchased from Wako Pure Chemical Industries, Ltd. 4,4-dibromobiphenyl (>98.0%), chlorodiphenylphosphine (>97.0%), and hexafluoroacetylacetone (>95.0%) were purchased from Tokyo Chemicals Industries Co. Ltd.

### *3.2.2 Apparatus*

ESI-MS was measured on a JEOL JMS-T100LP. Elemental analyses were performed by an Exter Analytical CE440. Infrared spectra were recorded on a JASCO FT/IR-4600 spectrometer. XRD patterns were characterized by a RIGAKU X-ray diffractometer RINT 2200. ICP-AES were obtained by using a Shimadzu ICPE-9000.

### *3.2.3 Synthesis*

The general synthetic method of the precursors and the coordination polymers are shown in Fig. 3.2 [21]. The synthetic procedures of the linker ligand, the precursor complexes, and the coordination polymers are based on the previous reports. Dpbp,

$Yb(hfa)_3(H_2O)_2$, $Gd(hfa)_3(H_2O)_2$, and the coordination polymers were synthesized also following the procedures previously reported. The procedure for the synthesis of $[Yb_xGd_{1-x}(hfa)_3dpbp]_n$ ($x$ = 1, 0.5, 0.1, 0.05, 0.01, 0.005, 0.001) coordination polymers is as follows: required amount of $Yb(hfa)_3(H_2O)_2$ and $Gd(hfa)_3(H_2O)_2$ (total 1 mmol) were dissolved in 10 mL of methanol and stirred for 5 min. Dpbp ligand (1 mmol, 0.56 g) dispersed in 20 mL of methanol was added subsequently and stirred for 5 h. The resulting white precipitate was vacuum filtered, washed with methanol, and dried in vacuum.

$[Yb_xGd_{1-x}(hfa)_3dpbp]_n$ coordination polymers were identified by infrared spectroscopy (IR), electron spray ionization mass spectroscopy (ESI-MS), and elemental analysis. Yb(III) ion concentration in each of the coordination polymers was determined by inductively coupled plasma atomic emission spectroscopy (ICP-AES). Diluted 1000 ppm standards of Yb and Gd nitrate aqueous solution were used as calibration standards. Calibration curves were made from three emission wavelengths of Yb (289.138, 328.937, 369.419 nm) and Gd (335.047, 336.223, 342.47 nm). The concentration of each coordination polymers were determined from the emission intensity of the three Yb and the three Gd wavelengths.

$[Yb_{0.001}Gd_{0.999}(hfa)_3dpbp]_n$: Yield: 1.21 g (91%, monomer). Selected IR (KBr, $cm^{-1}$): 3141 (w, arC–H), 3067 (w, arC–H), 1654 (st, C=O), 1599 (m, arC=C), 1551 (m, arC=C), 1527 (st, arC=C), 1498 (st, arC=C), 1439 (m, arP–C), 1254 (st, P=O), 1100–1200 (st, C–F). ESI-MS: m/z = 1126.06 $[Gd(hfa)_2dpbp]^+$, 1142.07 $[Yb(hfa)_2dpbp]^+$, 2459.10 $[Gd(hfa)_3dpbp][Gd(hfa)_2dpbp]^+$, 3792.13 $[Gd(hfa)_3dpbp]_2[Gd(hfa)_2dpbp]^+$. Elemental analysis calculated for $[C_{51}H_{31}F_{18}O_8P_2Yb_{0.001}Gd_{0.999}]_n$: C, 45.95%, H, 2.34%. Found: C, 46.18%, H, 2.23%. Yb and Gd ratio from ICP-AES: Yb:Gd = 0.001517:0.9915 (calcd. 0.002:0.998).

$[Yb_{0.005}Gd_{0.995}(hfa)_3dpbp]_n$: Yield: 1.23 g (92%, monomer). Selected IR (KBr, $cm^{-1}$): 3141 (w, arC–H), 3067 (w, arC–H), 1654 (st, C=O), 1599 (m, arC=C), 1551 (m, arC=C), 1527 (st, arC=C), 1498 (st, arC=C), 1439 (m, arP–C), 1254 (st, P=O), 1100–1200 (st, C–F). ESI-MS: m/z = 1126.07 $[Gd(hfa)_2dpbp]^+$, 1142.08 $[Yb(hfa)_2dpbp]^+$, 2459.09 $[Gd(hfa)_3dpbp][Gd(hfa)_2dpbp]^+$, 3791.10 $[Gd(hfa)_3dpbp]_2[Gd(hfa)_2dpbp]^+$. Elemental analysis calculated for $[C_{51}H_{31}F_{18}O_8P_2Yb_{0.005}Gd_{0.995}]_n$: C, 45.95%, H, 2.34%. Found: C, 46.15%, H, 2.23%. Yb and Gd ratio from ICP-AES: Yb:Gd = 0.00278:0.9944 (calcd. 0.003:0.997).

$[Yb_{0.01}Gd_{0.99}(hfa)_3dpbp]_n$: Yield: 1.25 g (93%, monomer). Selected IR (KBr, $cm^{-1}$): 3141 (w, arC–H), 3067 (w, arC–H), 1654 (st, C=O), 1599 (m, arC=C), 1551 (m, arC=C), 1527 (st, arC=C), 1498 (st, arC=C), 1439 (m, arP–C), 1254 (st, P=O), 1100–1200 (st, C–F). ESI-MS: m/z = 1126.04 $[Gd(hfa)_2dpbp]^+$, 1142.05 $[Yb(hfa)_2dpbp]^+$, 2459.06 $[Gd(hfa)_3dpbp][Gd(hfa)_2dpbp]^+$, 3791.7 $[Gd(hfa)_3dpbp]_2[Gd(hfa)_2dpbp]^+$. Elemental analysis calculated for $[C_{51}H_{31}F_{18}O_8P_2Yb_{0.01}Gd_{0.99}]_n$: C, 45.95%, H, 2.34%. Found: C, 46.07%, H, 2.22%. Yb and Gd ratio from ICP-AES: Yb:Gd = 0.005969:1.0175 (calcd. 0.006:0.994).

$[Yb_{0.05}Gd_{0.95}(hfa)_3dpbp]_n$: Yield: 1.22 g (92%, monomer). Selected IR (KBr, $cm^{-1}$): 3141 (w, arC–H), 3067 (w, arC–H), 1654 (st, C=O), 1599 (m, arC=C), 1551 (m, arC = C), 1527 (st, arC = C), 1498 (st, arC = C), 1439 (m, arP-C), 1254 (st, P=O), 1100–1200 (st, C–F).. ESI-MS: m/z = 1126.04 $[Gd(hfa)_2dpbp]^+$, 1142.05 $[Yb(hfa)_2dpbp]^+$, 2459.06 $[Gd(hfa)_3dpbp][Gd(hfa)_2dpbp]^+$, 3791.07 $[Gd(hfa)_3dpbp]_2[Gd(hfa)_2dpbp]^+$. Elemental analysis calculated for $[C_{51}H_{31}F_{18}O_8P_2Yb_{0.05}Gd_{0.95}]_n$: C, 45.93%, H, 2.34%. Found: C, 46.19%, H, 2.24%. Yb and Gd ratio from ICP-AES: Yb:Gd = 0.03498:0.9605 (calcd. 0.04:0.96).
$[Yb_{0.1}Gd_{0.9}(hfa)_3dpbp]_n$: Yield: 1.18 g (89%, monomer). Selected IR (KBr, $cm^{-1}$): 3141 (w, arC–H), 3067 (w, arC–H), 1654 (st, C=O), 1599 (m, arC=C), 1551 (m, arC=C), 1527 (st, arC=C), 1498 (st, arC=C), 1439 (m, arP–C), 1254 (st, P=O), 1100–1200 (st, C–F). ESI-MS: m/z = 1126.04 $[Gd(hfa)_2dpbp]^+$, 1142.05 $[Yb(hfa)_2dpbp]^+$, 2459.06 $[Gd(hfa)_3dpbp][Gd(hfa)_2dpbp]^+$, 2475.09 $[Gd(hfa)_3dpbp][Yb(hfa)_2dpbp]^+$ and $[Yb(hfa)_3dpbp][Gd(hfa)_2dpbp]^+$, 3791.08 $[Gd(hfa)_3dpbp]_2[Gd(hfa)_2dpbp]^+$. Elemental analysis calculated for $[C_{51}H_{31}F_{18}O_8P_2Yb_{0.1}Gd_{0.9}]_n$: C, 45.90%, H, 2.34%. Found: C, 46.11%, H, 2.22%. Yb and Gd ratio from ICP-AES: Yb:Gd = 0.07413:0.9383 (calcd. 0.07:0.93).
$[Yb_{0.5}Gd_{0.5}(hfa)_3dpbp]_n$: Yield: 1.21 g (91%, monomer). Selected IR (KBr, $cm^{-1}$): 3141 (w, arC–H), 3067 (w, arC–H), 1654 (st, C=O), 1599 (m, arC=C), 1551 (m, arC=C), 1527 (st, arC=C), 1498 (st, arC=C), 1439 (m, arP–C), 1254 (st, P=O), 1100–1200 (st, C–F). ESI-MS: m/z = 1126.05 $[Gd(hfa)_2dpbp]^+$, 1142.06 $[Yb(hfa)_2dpbp]^+$, 2459.08 $[Gd(hfa)_3dpbp][Gd(hfa)_2dpbp]^+$, 2475.09 $[Gd(hfa)_3dpbp][Yb(hfa)_2dpbp]^+$ and $[Yb(hfa)_3dpbp][Gd(hfa)_2dpbp]^+$, 3791.09 $[Gd(hfa)_3dpbp]_2[Gd(hfa)_2dpbp]^+$, 3808.12 $[Yb(hfa)_3dpbp][Gd(hfa)_3dpbp][Gd(hfa)_2dpbp]^+$, 3822.13 $[Gd(hfa)_3dpbp][Yb(hfa)_3dpbp][Yb(hfa)_2dpbp]^+$, 3839.15 $[Yb(hfa)_3dpbp]_2[Yb(hfa)_2dpbp]^+$. Elemental analysis calculated for $[C_{51}H_{31}F_{18}O_8P_2Yb_{0.5}Gd_{0.5}]_n$: C, 45.68%, H, 2.33%. Found: C, 45.87%, H, 2.22%. Yb and Gd ratio from ICP-AES: Yb:Gd = 0.4001:0.5965 (calcd. 0.4:0.6).
$[Yb_1(hfa)_3dpbp]_n$: Yield: 1.25 g (93%, monomer). Selected IR (KBr, $cm^{-1}$): 3141 (w, arC–H), 3067 (w, arC–H), 1654 (st, C=O), 1599 (m, arC=C), 1551 (m, arC=C), 1527 (st, arC=C), 1498 (st, arC=C), 1439 (m, arP–C), 1254 (st, P=O), 1100–1200 (st, C–F). ESI-MS: m/z = 1142.06 $[Yb(hfa)_2dpbp]^+$, 2491.10 $[Yb(hfa)_3dpbp][Yb(hfa)_2dpbp]^+$, 3839.13 $[Yb(hfa)_3dpbp]_2[Yb(hfa)_2dpbp]^+$. Elemental analysis calculated for $[C_{51}H_{31}F_{18}O_8P_2Yb]_n$: C, 45.42%, H, 2.32%. Found: C, 45.48%, H, 2.20%. Yb and Gd ratio from ICP-AES: Yb:Gd = 0.9917:0 (calcd. 1:0).

### 3.2.4 Spectroscopy

Absorption spectra were obtained by using a JASCO V-670 spectrometer. Emission spectra and lifetimes were measured using a Horiba/Jobin-Yvon FluoroLog-3 ps spectrofluorometer. The spectrofluorometer was equipped with a Hamamatsu R5509-43 photomultiplier for detecting NIR emission, and SpectraLED 356 nm light source was used as the excitation light source for the lifetime measurements.

### 3.2.5 *Crystallography*

Single crystals of $[Yb_1(hfa)_3dpbp]_n$ coordination polymer were obtained by liquid-liquid diffusion of dpbp dichloromethane solution and $Yb(hfa)_3(H_2O)_2$ methanol solution. One of the obtained colorless block crystals was mounted on a glass fiber by epoxy resin glue and cooled to −150 °C. All measurements were made using a Rigaku RAXIS RAPID imaging plate area detector with graphite-monochromated MoKα radiation. Corrections for decay and Lorentz-polarization effects were made using a spherical absorption correction, solved by direct methods, and expanded using Fourier techniques. Non-hydrogen atoms were refined anisotropically. Hydrogen atoms were refined using the riding model. The final cycle of full-matrix least-squares refinement was based on observed reflections and variable parameters. All calculations were performed using a CrystalStructure crystallographic software package. The CIF was checked using the checkCIF/PLATON service. The CIF was deposited to The Cambridge Crystallographic Data Center (CCDC) with the deposition number CCDC-1046310 ($[Yb_1(hfa)_3dpbp]_n$). The data can be obtained free of charge from CCDC via www.ccdc.cam.ac.uk/data_request/cif.

## 3.3 Theoretical Methods

### 3.3.1 *Rate Constants*

The rate constants for the radiative process (electric dipole and magnetic dipole transitions), as well as energy transfer of various interactions, are provided in Sect. 2.2. Here, the relevant equations are listed. The radiative rate constants are [22–24]:

$$A_{\text{rad(ED)}} = \frac{2\pi v_{\text{em}}^2 e^2}{\varepsilon_0 c^3 m} n^2 f_{\text{ED}}, \tag{2.24}$$

$$A_{\text{rad(MD)}} = \frac{2\pi v_{\text{em}}^2 e^2}{\varepsilon_0 c^3 m} n^2 f_{\text{MD}}, \tag{2.25}$$

where

$$f_{\text{ED}} = \frac{8\pi^2 m_e}{3h} \frac{v_{\text{abs}}}{2J+1} \chi_{\text{abs}} \sum_{k=2,4,6} \Omega_k \left| \left\langle l^N SLJ_{\text{mixed}} \right\| U^{(k)} \left\| l^N S'L'J'_{\text{mixed}} \right\rangle \right|^2, \tag{2.13}$$

$$f_{\mathrm{MD}} = \frac{h\nu_{\mathrm{abs}}}{6m_e c^2}\frac{n}{2J+1}\left|\langle l^N SLJ_{\mathrm{mixed}}\| L + gS \| l^N S'L'J'_{\mathrm{mixed}}\rangle\right|^2. \tag{2.18}$$

The energy transfer rate constants are [25, 26]:

$$W_{\mathrm{ET(dd)}} = \frac{1}{(2J_{\mathrm{a}}+1)(2J_{\mathrm{b}}+1)}\frac{2}{3}\frac{4\pi^2}{h}\frac{e^4}{R^6}\left[\sum_{k=2,4,6}\Omega_k^A\left|\langle l^N SLJ\|U^{(k)}\|l^N S'L'J'\rangle\right|^2\right] \times\left[\sum_{k=2,4,6}\Omega_k^B\left|\langle l^N SLJ\|U^{(k)}\|l^N S'L'J'\rangle\right|^2\right]F, \tag{2.28}$$

$$W_{\mathrm{ET(dq)}} = \frac{1}{(2J_{\mathrm{a}}+1)(2J_{\mathrm{b}}+1)}\frac{4\pi^2}{h}\frac{e^4}{R^8}\left[\sum_{k=2,4,6}\Omega_k^A\left|\langle l^N SLJ\|U^{(k)}\|l^N S'L'J'\rangle\right|^2\right] \times\left[\left|\langle 4\mathrm{f}r_B^2 4\mathrm{f}\rangle\right|^2\left|\langle l\|C^{(2)}\|l\rangle\right|^2\left|\langle l^N SLJ\|U^{(2)}\|l^N S'L'J'\rangle\right|^2\right]F, \tag{2.29}$$

$$W_{\mathrm{ET(qq)}} = \frac{1}{(2J_{\mathrm{a}}+1)(2J_{\mathrm{b}}+1)}\frac{14}{5}\frac{4\pi^2}{h}\frac{e^4}{R^{10}}\left|\langle l\|C^{(2)}\|l\rangle\right|^4\left[\left|\langle 4\mathrm{f}\|r_A^2\|4\mathrm{f}\rangle\right|^2\left|\langle l^N SLJ\|U^{(2)}\|l^N S'L'J'\rangle\right|^2\right] \times\left[\left|\langle 4\mathrm{f}\|r_B^2\|4\mathrm{f}\rangle\right|^2\left|\langle l^N SLJ\|U^{(2)}\|l^N S'L'J'\rangle\right|^2\right]F. \tag{2.30}$$

In general, these rate constants require several fitting procedures with experimental data to calculate the values. However, below is an explanation how this is not necessary for this investigation.

Ce(III) ion can be considered as a one-electron system (only $^2F_J$ term) as far as the 4f-electrons are concerned. Since the terms for lanthanides with $N$ and 14-$N$ electrons are the same (see Chap. 2), Yb(III) ions can also be considered as a one-electron system (one can interpret this by considering that a Yb(III) ion contain one hole). In such case, electrostatic coupling does not exist, and only spin-orbit coupling contribute to the splitting of $4f^N$-states. Furthermore, spin-orbit coupling mixes states of the same $J$, which does not exist for Ce(III) and Yb(III) ions. Therefore, the wavefunctions of these ions are not mixed in the intermediate coupling scheme. The reduced matrix element in Eqs. (2.13), (2.18), (2.28), and (2.29) can now be directly calculated without accounting for mixed wavefunctions, and therefore Eqs. (2.24) and (2.25) as well.

Yb(III) ions only have one excited state (i.e., $^2F_{5/2}$). Since $f_{\mathrm{tot}} = f_{\mathrm{ED}} + f_{\mathrm{MD}}$ (Eq. 2.22), the value of the only $f_{\mathrm{ED}}$ can be calculated simply from oscillator strength measured from absorption spectra. The entire sum $\sum_{k=2,4,6}\Omega_k\left|\langle l^N SLJ_{\mathrm{mixed}}\|U^{(k)}\|l^N S'L'J'_{\mathrm{mixed}}\rangle\right|^2$ can be calculated from Eq. (2.13). The low solubility of $[\mathrm{Yb(hfa)_3dpbp}]_n$ coordination polymers rendered the oscillator strength of Yb(III) ions undetermined, so instead a typical oscillator strength $f_{\mathrm{tot}} = 3 \times 10^{-6}$ was used [25, 26]. Therefore, the discussion is restricted to the qualitative side of the theoretical results and not the quantitative side.

Phonon-assisted ET relaxation (PAETR) is an additional relaxation process that occurs when two luminescent centers are close in the distance. The principle of this type of ET is the same as the phonon-assisted ET, where one luminescent center in its excited state transfers its energy to a ground state luminescent center with different excited state energy, assisted with a phonon(s) to compensate for the difference in energy. F. Auzel proposed in 2002 that phonon-assisted ET can also occur as a relaxation process when one luminescent center transfers directly to the phonon state of the neighboring luminescent center (hence can be considered as being "fully phonon-assisted") [15]. The rate constant of phonon-assisted ET $W_{\mathrm{PAET}}$ is given by the following equation provided by Miyakawa and Dexter [16]:

$$W_{\mathrm{PAET}} = W_{\mathrm{tot}}\, e^{-\frac{\Delta E}{\hbar\omega}\ln\frac{N}{se(n+1)\left(1+{}^{s_1}/{s_2}\right)}}. \tag{3.1}$$

where $W_{\mathrm{tot}}$ is phonon-unassisted ET rate constant $\left(= W_{\mathrm{ET(dd)}} + W_{\mathrm{ET(qd)}} + W_{\mathrm{ET(dq)}} + W_{\mathrm{ET(qq)}}\right)$, $\Delta E$ is energy gap between the excited state of two Ln(III) ions, $\hbar\omega$ is the phonon energy relevant to the assist, $s$ is the Huang-Rhys parameter ($s_1$ and $s_2$ refers to donor and acceptor), $n$ is number of phonon at a given temperature, $N$ is the multi-phonon order given by the integer part of $N = \Delta E/\hbar\omega$. In the case of PAETR, F. Auzel described that $\Delta E$ is the energy gap between ground and excited state of the donor ($= \upsilon_{\mathrm{em}}$). When the donor and the acceptor are the same species, $s_1$ and $s_2$ are identical and Eq. (3.1) can be modified for PAETR and becomes:

$$W_{\mathrm{PAETR}} = W_{\mathrm{tot}}\, e^{-\frac{\Delta E}{\hbar\omega}\ln\frac{N}{2se(n+1)}}. \tag{3.2}$$

$s$ is provided by the Stokes shift between absorption and emission spectra $\Delta$, which for lanthanides are typically in the order of $1 \times 10^{-2}$:

$$s = \frac{\Delta}{2\times\hbar\omega}. \tag{3.3}$$

$n$ can be derived from Boltzmann distribution:

$$n = \frac{1}{e^{\frac{\hbar\omega}{k_B T}}-1}. \tag{3.4}$$

The C–H stretching mode of the alpha position of hfa ligands (3141 $\mathrm{cm}^{-1}$, see Sect. 3.2.3) was considered as the relevant phonon for this process since it has been shown in previous reports thsat this vibration is important in suppressing vibrational relaxation of lanthanide complexes. The various constants (experimental and calculated) are summarized in Table 3.1.

**Table 3.1** Values for various constants needed for rate equation calculation

| Constant | Value | Constant | Value |
|---|---|---|---|
| $v_{abs}$ | $1.039 \times 10^4$ cm$^{-1}$ | $f_{ED}$ | $2.729 \times 10^{-6}$ |
| $v_{em}$ | $1.007 \times 10^4$ cm$^{-1}$ | $\sum_{k=2,4,6} \Omega_k \left|\langle l^N SLJ \| U^{(k)} \| l^N S'L'J' \rangle\right|^2$ | $9.652 \times 10^{-21}$ cm$^2$ |
| $\hbar\gamma_D$ | $5.845 \times 10^2$ cm$^{-1}$ | $A_{rad(MD)}$ | $4.395 \times 10^1$ s$^{-1}$ |
| $\hbar\gamma_A$ | $6.794 \times 10^2$ cm$^{-1}$ | $A_{rad(ED)}$ | $4.420 \times 10^2$ s$^{-1}$ |
| $A_{tot}$ | $4.611 \times 10^4$ s$^{-1}$ | $A_{nrad}$ | $4.562 \times 10^4$ s$^{-1}$ |
| $\chi_{abs}$ | 2.007 | $W_{ET(dd)}$ | $9.688 \times 10^8 \times R^{-6}$ [s$^{-1}$] |
| $\Delta$ | $3.178 \times 10^2$ cm$^{-1}$ | $W_{ET(dq)}$ | $1.289 \times 10^{11} \times R^{-8}$ [s$^{-1}$] |
| $F$ | $4.804 \times 10^{-4}$ cm | $W_{ET(qq)}$ | $3.201 \times 10^{13} \times R^{-10}$ [s$^{-1}$] |
| $f_{total}$ | $3 \times 10^{-6}$ | $\hbar\omega$ | 3141 cm$^{-1}$ |
| $\langle l^N SLJ \| L + gS \| l^N SL(J-1) \rangle$ | 1.856 | $s$ | 0.096 |
| $\langle l^N SL \| U^{(k)} \| l^N SL' \rangle$ | – 1.000 ($k = 2, 4, 6$) | $n$ | $2.869 \times 10^{-7}$ |
| $\langle l^N SLJ \| U^{(k)} \| l^N SL'J' \rangle$ | 0.3499 ($k = 2$)<br>0.6389 ($k = 4$)<br>0.9258 ($k = 6$) | $N$ | 3 |
| $\langle l \| C^{(2)} \| l \rangle$ | – 1.366 | $W_{PAETR}$ | $W_{tot} \times 3.670 \times 10^{-3}$[s$^{-1}$] |
| $f_{MD}$ | $2.713 \times 10^{-7}$ | | |

### 3.3.2 Rate Equations

The concept of the theory and the program to automatically generate rate equations are briefly explained in this section. The flow chart of the calculation method is summarized in Fig. 3.3. As mentioned in the main text, "active sphere" considered for calculation consists of 135 Ln(III) ions, which is approximately 40 Å in diameter (Fig. 3.4). In this active sphere, the minimum concentration is 0.01 where only the center Ln(III) ion is Yb(III) ion (red dot), and the rest is considered to be Gd(III) ions, and the maximum concentration is unity where all of the sites are Yb(III) ions. With reasonable excitation power (under 0.02 W $mm^{-2}$), multiple excitations in the same sphere are neglected [27], thus for this calculation, initial excitation can be considered to begin at the center Yb(III) ion. With these considerations, this active sphere can be modeled as a system of inhomogeneous linear differential equations, and its matrix form (see Sect. 2.3) is given by the following equation [28]:

$$\frac{\mathrm{d}\boldsymbol{X}(t)}{\mathrm{d}t} = \boldsymbol{A}\boldsymbol{X}(t) + \boldsymbol{J}(t), \tag{2.36}$$

where $\boldsymbol{X}(t)$ is the vector of population density, and each of the elements represents each excited Yb(III) ion, and $\boldsymbol{J}(t)$ is the input function of the system. $\boldsymbol{A}$ is the "interaction matrix," a 135 × 135 square matrix consisting of rate constants that models

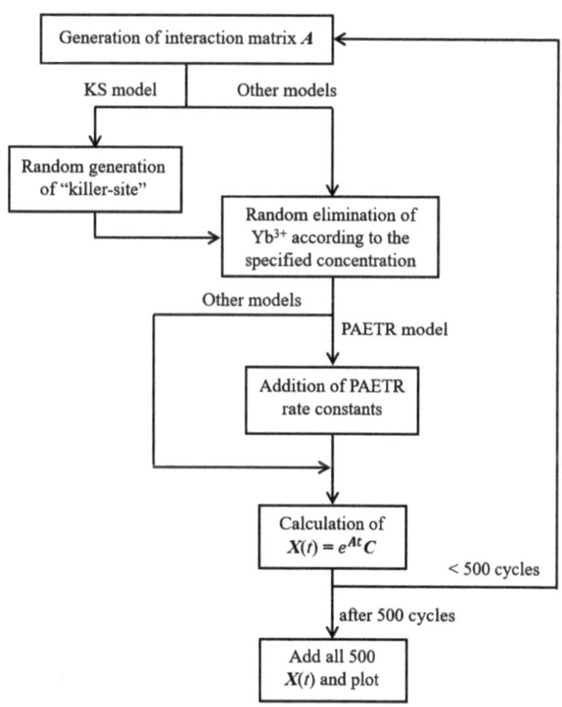


**Fig. 3.3** The flow chart of the rate equation calculation

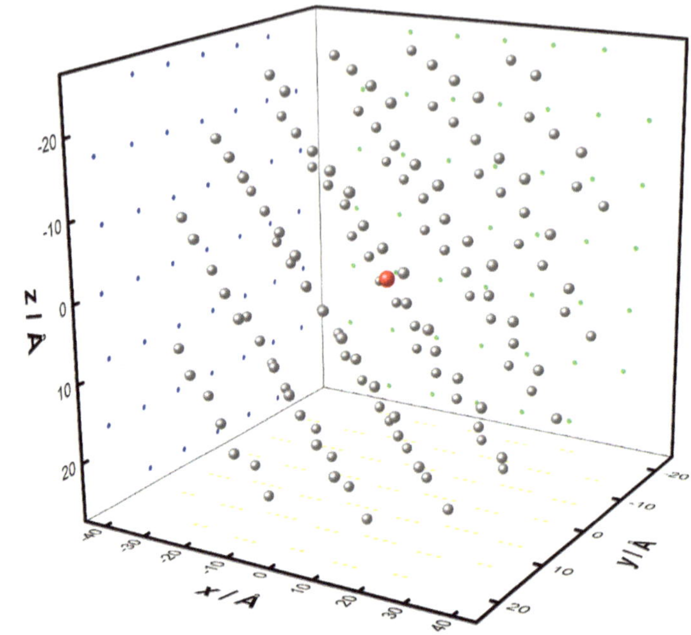


**Fig. 3.4** "Active sphere" of the calculation consisting of 135 Ln(III) ions. The grey spheres represent Yb(III) ions. The red sphere represents the center Yb(III) ion. The coordinates were taken from single crystal X-ray analysis. The diameter is approximately 40 Å

the interaction between Yb(III) ions as well as the fundamental decay rate constant of each Yb(III) ion:

$$\boldsymbol{A} = \begin{bmatrix} -(A_{\mathrm{tot}} + W_{\mathrm{tot}-67}) & W_{\mathrm{ET}-66,-67} & \cdots & W_{\mathrm{ET}0,-67} & \cdots & \cdots & W_{\mathrm{ET}67,-67} \\ W_{\mathrm{ET}-67,-66} & \ddots & \ddots & \vdots & & & \vdots \\ \vdots & \ddots & -(A_{\mathrm{tot}} + W_{\mathrm{tot}-1}) & W_{\mathrm{ET}0,-1} & & \ddots & \vdots \\ W_{\mathrm{ET}-67,0} & \cdots & W_{\mathrm{ET}-1,0} & -(A_{\mathrm{tot}} + W_{\mathrm{tot}0}) & W_{\mathrm{ET}1,0} & \cdots & W_{\mathrm{ET}67,0} \\ \vdots & & & W_{\mathrm{ET}0,1} & -(A_{\mathrm{tot}} + W_{\mathrm{tot}1}) & \ddots & \vdots \\ \vdots & & & \vdots & \ddots & \ddots & W_{\mathrm{ET}67,66} \\ W_{\mathrm{ET}-67,67} & \cdots & \cdots & W_{\mathrm{ET}0,67} & \cdots & W_{\mathrm{ET}66,67} & -(A_{\mathrm{tot}} + W_{\mathrm{tot}67}) \end{bmatrix}, \tag{3.5}$$

where $A_{\mathrm{tot}} = A_{\mathrm{rad}} + A_{\mathrm{nrad}}$, $W_{\mathrm{ET}i,j}$ is the ET rate constant between $i$th and $j$th Yb(III) ions, and $W_{\mathrm{tot}i}$ is the total rate constant $\left(= W_{\mathrm{ET}i,1} + W_{\mathrm{ET}i,2} + \cdots + W_{\mathrm{ET}i,135}\right)$ of the ET from the $i$th Yb(III) ion. The diagonal elements represent decay rate of each Yb(III) ion, and the off-diagonal elements represent the ET rate constant between Yb(III) ions. Delta function is used for the center Yb(III) ion as the input function $\boldsymbol{J}(t)$ for the calculation of emission decay profile, and thus Eq. (2.36) can be solved to give:

$$X(t) = \exp(At)B, \tag{2.43}$$

where $\boldsymbol{B}$ is the vector of the initial population density for all Yb(III) ions, which is zero for all elements except for the element corresponding to the center Yb(III) ion (68th element). $\exp(\boldsymbol{A}t)$ is the matrix exponential of the interaction matrix. Diagonalization of $\exp(\boldsymbol{A}t)$ in Eq. (2.43) solves for $\boldsymbol{X}(t)$. The calculated $\boldsymbol{X}(t)$ provides the population density decay profiles of the active sphere. This procedure applies for all of the calculations in this section.

The following part of this section explains the modification of the interaction matrix $\boldsymbol{A}$ (Eq. (3.5)) corresponding to the specified concentration and the concentration quenching model. In order to account for the inhomogeneity of the distribution of Yb(III) ions, the row and column were changed to zero for those corresponding to the Yb(III) ions randomly specified by uniformly distributed random integer generator. Then all the elements of each column except the diagonal element were added, and the diagonal element was replaced with the sum along with $A_{\text{rad}}$ and $A_{\text{nrad}}$. This efficiently erases all the rate constants existent in the system that involves the specified Yb(III) ion.

In the killer-site (KS) model, an abstract form of "killer-site" was introduced. Identifying the specific type of killer-site is difficult and is beyond the scope of this work. For the sake of discussion, non-radiative impurity-like killer-site was considered, which is acting as strong quenching center (rate constant of the ET to this site is 1000 times larger than the YbYbET). The concentration of the killer-site is decided randomly from 0 to 1 (corresponds to approx. 0.007 of all Ln(III) ion sites), and the distribution is also decided randomly, both using uniformly distributed random integer generator. The column corresponding to the killer-site chosen by the uniformly random integer generator is set to zero, and the diagonal element is changed to $A_{\text{nrad}}$ only, which removes all of the ET back to the Yb(III) ions and decays without emission. The row corresponding to the killer-site is multiplied by 1000 to model the killer-site as strong quenching center.

In the PAETR model, an additional relaxation rate is described by Eq. (3.2). $W_{\text{tot}}$ is the total ET rate constant for individual Yb(III) ion, and this is the origin of concentration dependence. The interaction matrix $\boldsymbol{A}$ is modified for PAETR model by adding the exponential part of Eq. (3.2) multiplied with $W_{\text{tot}}$ of individual Yb(III) ion to the diagonal elements.

Above method describes single calculation for a single active sphere. Experimentally, the emission lifetime of the combination of multiple active spaces is observed, each with a different distribution of Yb(III) ions. To account for this inhomogeneity, 500 cycles of the calculation were performed and combined by adding each of the derived equations. The calculation was done in MATLAB R2015b, and the scripts are provided in Sect. 3.6.2.

## 3.4 Results and Discussion

### *3.4.1 Structure*

The crystal structure of $[Yb_1(hfa)_3dpbp]_n$ coordination polymer was determined by single-crystal X-ray analysis (Fig. 3.5a). $[Yb_1(hfa)_3dpbp]_n$ forms a polymer structure with each of the Yb(III) ions coordinated by two dpbp and three hfa ligands. As shown in Fig. 3.5b, the shortest pair of Yb(III) is between the neighboring polymer chains, separated by 10.4 Å. The structure is similar to that of the previously reported $[Eu(hfa)_3dpbp]_n$ [21]. Powder XRD analyses of $[Yb_xGd_{1-x}(hfa)_3dpbp]_n$ ($x = 1, 0.5,$ $0.1, 0.05, 0.01, 0.005, 0.001$) coordination polymers allows comparison of the periodic structures. As shown in Fig. 3.5c, the main peaks at around 10.2, 20.1, and 21.5 degrees of the polymers are within negligibly small shifts of 0.1 degrees, with similar FWHMs. The diffuse reflectance spectra, which is directly related to the absorption spectra, in the NIR region of $[Yb_1(hfa)_3dpbp]_n$ and $[Yb_{0.01}Gd_{0.99}(hfa)_3dpbp]_n$ were recorded to determine the dependence of concentration on the local coordination geometry around Yb(III) ions (Fig. 3.10, see Sect. 3.6.1). The emission and absorption spectral shapes of Ln(III) ions are known to be sensitive to the local coordination geometry around Ln(III) ion [29]. The resulting spectrum was not of high quality for $[Yb_{0.01}Gd_{0.99}(hfa)_3dpbp]_n$, and thus it was not possible to compare the spectral shape. The diffuse reflectance spectrum of $[Yb_{0.01}Gd_{0.99}(hfa)_3dpbp]_n$ took about 11 days to complete due to the small absorption coefficient of Yb(III) ions while that of $[Yb_1(hfa)_3dpbp]_n$ took only about 10 h. The resolution, sensitivity, correction, and the baseline might have changed during such extreme time duration differences. Nonetheless, the lack of notable change in the powder XRD patterns shows that the structures of $[Yb_xGd_{1-x}(hfa)_3dpbp]_n$ coordination polymers are identical.

### *3.4.2 Photophysical Properties*

In order to consider the effect of self-trapping, emission spectra of $[Yb_xGd_{1-x}(hfa)_3dpbp]_n$ coordination polymers in the NIR region were measured in powder state (excitation at 356 nm). The emission spectra of the polymers are shown in Fig. 3.6a. Multiple emission peaks from 900 to 1100 nm attributed to the ${}^2F_{5/2} \rightarrow {}^2F_{7/2}$ transition of Yb(III) ions with Stark splitting were observed. The height of the peaks at a shorter wavelength decreased with increasing concentration of Yb(III) ions. Figure 3.6b shows the comparison of diffuse reflectance and emission spectra of $[Yb_{0.01}Gd_{0.99}(hfa)_3dpbp]_n$. The longer wavelength region of the diffuse reflectance spectrum overlaps with the shorter wavelength region of the emission spectrum. The overlap leads to the reabsorption of emission in the shorter wavelength region when the Yb(III) ion concentration is high. This result confirms the existence of self-trapping in $[Yb_xGd_{1-x}(hfa)_3dpbp]_n$ coordination polymers.

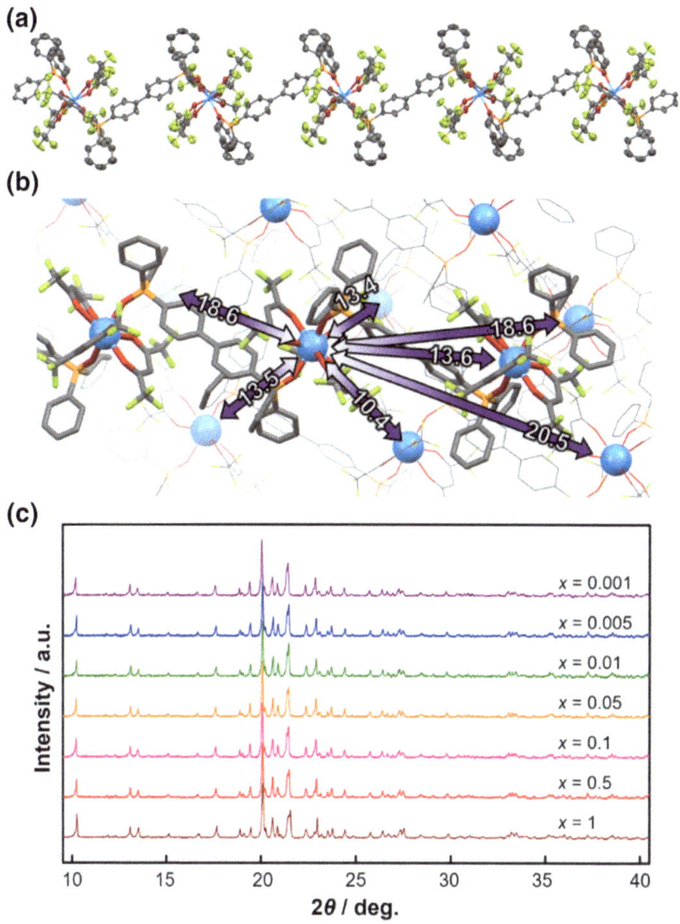


**Fig. 3.5** **a** Molecular structure of the $[Yb_1(hfa)_3dpbp]_n$ coordination polymer, showing 50% probability displacement ellipsoids. Hydrogen atoms are omitted for clarity. Blue: Yb, red: O, gray: C, yellow: F, and orange: P. **b** Packing structure of $[Yb_1(hfa)_3dpbp]_n$ polymer chains and the distance between Yb(III) ions in angstroms (Å). **c** Powder XRD patterns of $[Yb_xGd_{1-x}(hfa)_3dpbp]_n$. Reprinted from ref [44] with permission from John Wiley & Sons LTD

Emission lifetimes provide information on the mechanism of concentration quenching. Emission decay profiles of $[Yb_xGd_{1-x}(hfa)_3dpbp]_n$ coordination polymers excited at 356 nm and monitored at 1007 nm (Yb(III) ion emission) are shown in Fig. 3.7a. The emission decay profiles were single-exponential at all concentrations of Yb(III) ions in the polymers. The emission lifetimes remained constant at approximately 21.8 μs for $x = 0.001$ to $x = 0.01$, and gradually decreased for higher concentrations (Fig. 3.7b). The emission lifetimes ultimately decreased to 19.2 μs for $x = 1$. According to the previous reports, self-trapping leads to an increase in emission lifetimes, while self-quenching leads to a decrease in emission lifetimes

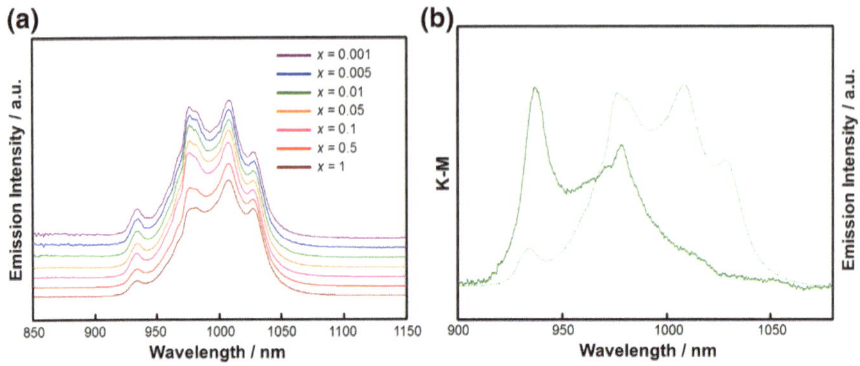


**Fig. 3.6** **a** Emission spectra of $[Yb_xGd_{1-x}(hfa)_3dpbp]_n$ coordination polymers ($\lambda_{EX}$ = 356 nm). The spectra were normalized at the 1025 nm peak, and offsets were made for clarity. **b** Diffuse reflectance spectrum (solid line) and emission spectrum (dotted line) of $[Yb_{0.01}Gd_{0.99}(hfa)_3dpbp]_n$. Reprinted from ref [44] with permission from John Wiley & Sons LTD

[12, 30]. An increase in the emission lifetimes was not observed, indicating the dominant effect of self-quenching over the effect of self-trapping. The weak effect of self-trapping is because the optical measurements were performed on the fine powder sample. The long minimum Yb(III)—Yb(III) distance of 10.4 Å typically means a limited-diffusion type of concentration quenching due to the slow energy migration. Since this type of concentration quenching leads to multi-exponential decay at high concentration in the presence of killer-sites, there is an inconsistency with the experimental result that showed single exponential decay. Therefore, interpreting the concentration quenching mechanism from the shape of the decay profiles is not simple. In the following section, the concentration quenching mechanisms is elucidated by qualitatively comparing the experimental lifetimes to those of theoretically calculated lifetimes.

### 3.4.3 *Theoretical Lifetimes*

The theoretical lifetimes were calculated for "no concentration quenching (NCQ)," "killer-site (KS)," and "phonon-assisted ET relaxation" (PAETR) models by the method provided in Sect. 3.3. Theoretical Methods. The theoretical emission decay profiles of $[Yb_xGd_{1-x}(hfa)_3dpbp]_n$ coordination polymers in the NCQ model for $x =$ 0.01, 0.05, 0.1, 0.5, and 1 are given in Fig. 3.8a. All emission lifetimes were identical both in the shape of the decay profile (being single-component) and the slope. The result indicates that the emission lifetimes do not depend on the concentration of Yb(III) ions (and thus the energy migration rate), in the absence of concentration quenching even when the distribution of Yb(III) ions is inhomogeneous.

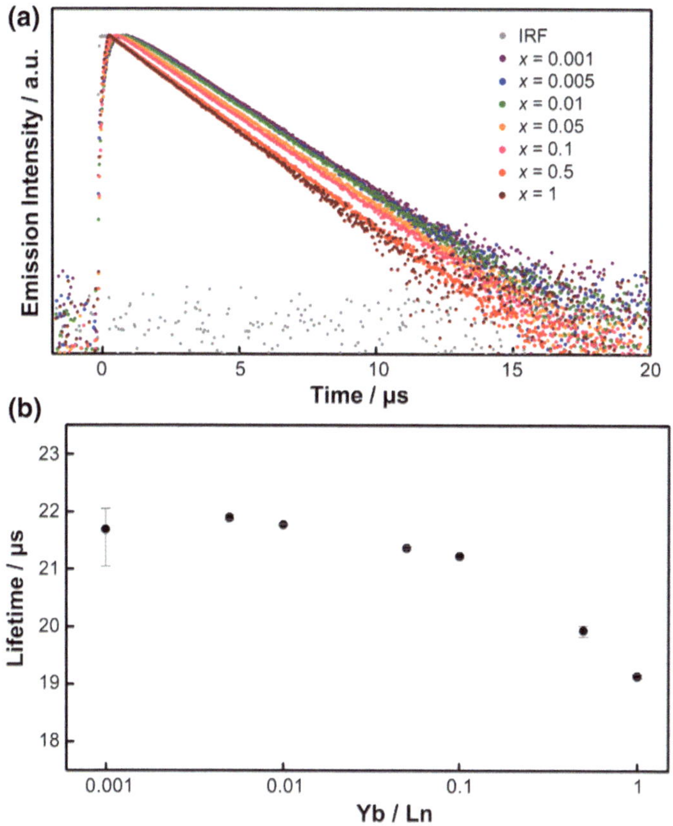


**Fig. 3.7** **a** Emission lifetime decay profile of $[Yb_xGd_{1-x}(hfa)_3dpbp]_n$ coordination polymers ($\lambda_{EX}$ = 356 nm, $\lambda_{EM}$ = 1007 nm). IRF: instrument response function. **b** Plot of emission lifetimes for all Yb(III) concentration in $[Yb_xGd_{1-x}(hfa)_3dpbp]_n$ coordination polymers. Reprinted from ref [44] with permission from John Wiley & Sons LTD

In the presence of concentration quenching, the coordination polymers exhibit different photophysical behavior between the KS model and the PAETR model. In the presence of a killer-site, the shape of the decay profile was multi-component for all concentration, even at $x = 0.01$ (Fig. 3.8b). The average lifetimes did not change significantly with concentration, contrary to what would generally be expected for this model. There are several points of the decay profiles of each concentration that crosses (Fig. 3.8b inset), indicating that the concentrations considered in this calculation are within the range of transition phase from a limited-diffusion to a fast-diffusion scheme. These features are not observed in the case of PAETR model as the decay profile remains single-component throughout all concentrations and gradually decreases with increasing concentration (Fig. 3.8c). The difference in the photophysical behavior between the two concentration quenching mechanisms arises from the degree of localization of the quenching. Concentration quenching in the KS model is

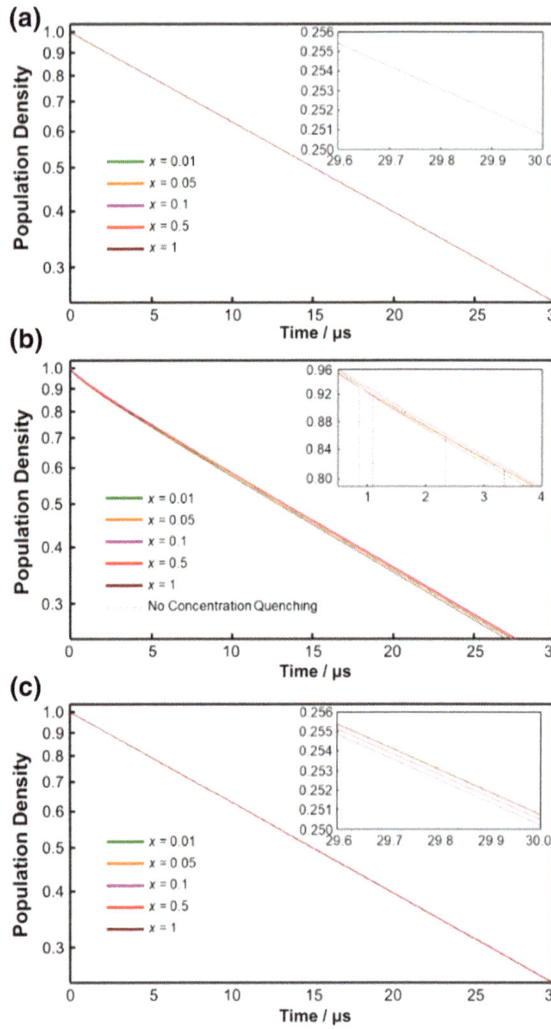


**Fig. 3.8** Theoretical time-evolution of the population density of the excited state Yb(III) ions of $[Yb_xGd_{1-x}(hfa)_3dpbp]_n$ coordination polymers for $x = 0.01$–1: **a** "No concentration quenching" (NCQ) model, **b** "killer-site" (KS) model, and **c** "phonon-assisted ET relaxation" (PAETR) model. The inset shows the close-up of the decay profiles. Time-evolution of the population density of the excited state Yb(III) ions is directly related to emission lifetimes. Modified and then reprinted from Fig. 3.5 (split into Fig. 3.8 and 3.9) in ref [44] with permission from John Wiley & Sons LTD

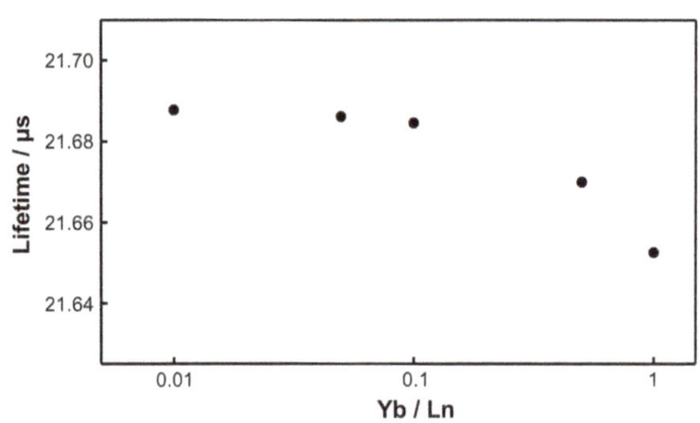


**Fig. 3.9** Plot of theoretical lifetimes for the PAETR model. Modified and then reprinted from Fig. 3.5 (split into Fig. 3.8 and 3.9) in ref [44] with permission from John Wiley & Sons LTD

a localized mechanism where the excited state energy is delivered to the killer-site via energy migration. In the case of $[Yb_xGd_{1-x}(hfa)_3dpbp]_n$ coordination polymers, the maximum ET rate for the central Yb(III) ion (red dot in Fig. 3.4) is 20,378 $s^{-1}$, which is approximately only the half of the fundamental decay rate constant ($A_{rad} + A_{nrad}$) of 46109 $s^{-1}$. This implies that energy migration does not diffuse to large spatial area and average out the lifetimes of individual Yb(III) ion. Depending on the position of the killer-site relative to the excited Yb(III) ion, the effect of the quenching can change drastically. In total, the emission from multiple spherical ensembles with the various position of killer-sites lead to the multi-component decay. On the other hand, concentration quenching in the PAETR model is a well-distributed effect since phonon-assisted ET relaxation occurs between all Yb(III) ions that are close. This results in the decay profile that is apparently single-component for all concentrations.

Of the three different models, the experimental emission decay profiles (Fig. 3.7a) are more consistent with the theoretical emission decay profiles for the PAETR model (Fig. 3.8c). Both show similar shape for the decrease in the emission lifetimes with concentration beyond $x = 0.01$ (Fig. 3.7b and 3.9), and the profiles are single-component at all concentrations. Meanwhile, the rate of decrease in the lifetimes is significantly smaller for the theoretical lifetime. This is likely because of two reasons: the increased electron-phonon coupling between Yb(III) ion pairs and the larger actual ET rate than the calculated values (see Sect. 3.6.3). Auzel [31, 32] and Vink [33] both reported that electron-phonon coupling strength itself increases with increased concentration of luminescent center. This could slightly increase in both $A_{nrad}$ and $W_{PAETR}$. Also, the theoretical lifetimes are calculated based on typical Yb(III) oscillator strength of $3 \times 10^{-6}$ but can range between $1 \times 10^{-6}$–$8 \times 10^{-6}$ according to the previous studies [34–38]. Since the ET rate constant was calculated from the oscillator strength (by deriving values for $\sum_{k=2,4,6} \Omega_k \left|\langle l^N SLJ \| U^{(k)} \| l^N S'L'J' \rangle\right|^2$ in Eqs. (2.23) and (2.24), deviation in the oscillator strength (and refractive index) can lead to deviations in the ET rate constant roughly from 0.04 to 36 times the calculated value (which also increases $W_{PAETR}$ around this order). In total, the deviations in

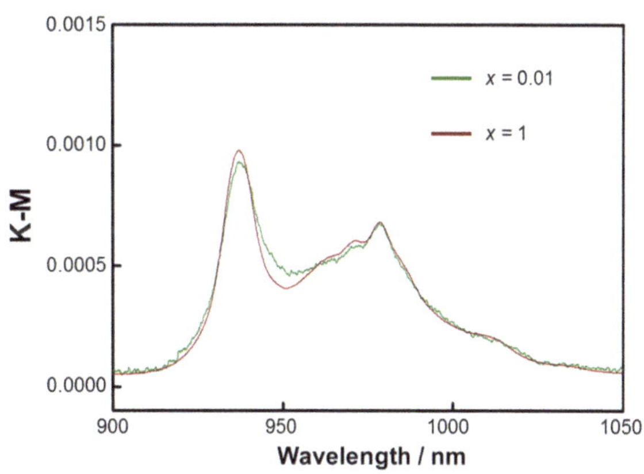


**Fig. 3.10** Diffuse reflectance spectra of $[Yb_1(hfa)_3dpbp]_n$ and $[Yb_{0.01}Gd_{0.99}(hfa)_3dpbp]_n$ in the NIR region

the quantitative values in this case are within expected order. The single component lifetime and the similar shape of the plot of the emission lifetimes qualitatively suggest that the concentration quenching mechanism in $[Yb_xGd_{1-x}(hfa)_3dpbp]_n$ coordination polymers is the phonon-assisted ET relaxation (Fig. 3.1).

The nonradiative rate constant, ET rate constant, and the model of concentration quenching are important factors that affect lifetimes, and they must be considered. As stated above, the ET rate constant is comparatively smaller than the fundamental decay rate constant in $[Yb_xGd_{1-x}(hfa)_3dpbp]_n$ coordination polymers. The maximum ET efficiency is 30.6%, and thus the excited Yb(III) ions are rather localized. The large nonradiative rate constant $A_{nrad}$ of 45623 $s^{-1}$ is due to the use of organic ligands with high-frequency vibration modes. In this sense, the small effect of concentration quenching in Ln(III) coordination polymers is not only because of the PAETR mechanism. The effect of concentration quenching is also lowered by the reduction in the energy migration due to the large fundamental decay rate constant of Ln(III) ions. This especially applies for NIR emitting Ln(III) ions because they are significantly more vulnerable to multi-phonon relaxation, as in the case of $[Yb(hfa)_3dpbp]_n$ coordination polymer. Therefore, an extra care needs to be taken before judging that concentration quenching is a small effect in lanthanide coordination polymers, because it may simply be due to the large nonradiative rate constant and not the fundamental property of lanthanide coordination polymers.

The fundamental difference between Ln(III) doped inorganic host and Ln(III) coordination polymers is likely the concentration quenching mechanism. Inorganic hosts usually show multi-component decay, suggesting the existence of killer-sites. On the other hand, various reported Ln(III) coordination polymers, including $[Yb(hfa)_3dpbp]_n$, at maximum concentration still show single-component emission lifetimes [39–43], implying that coordination polymers in general lack or have an extremely low concentration of killer-sites. Therefore, the concentration quenching for coordination polymers is suggested to be dominated by the PAETR, a small effect compared to strongly quenching killer-sites. Other than to lower the concentration of Ln(III) ions, the strategy to reduce the effect of concentration quenching in both types of the host is different. For inorganic hosts, this is reducing the concentration of killer-sites (high purity host). For Ln(III) coordination polymers, the general strategy to achieve high emission quantum yield in lanthanide complexes of eliminating phonon modes responsible for multi-phonon relaxation is also effective in reducing concentration quenching. Equation (3.2) indicates that smaller phonon energy $\hbar\omega$ leads to smaller PAETR rate constant $W_{PAETR}$. The results provides a new meaning in using Ln(III) coordination polymers as functional luminescent materials based on energy transfer.

## 3.5 Conclusion

The concentration quenching mechanism and the extent of its effect in $[Yb_xGd_{1-x}(hfa)_3dpbp]_n$ ($x = 0.001, 0.005, 0.01, 0.05, 0.1, 0.5, 1$) coordination polymers were investigated. Emission spectra revealed that the shorter wavelength region of the Yb(III) emission is deformed with increasing concentration of Yb(III) ions, indicating self-trapping. Emission lifetimes were single-component at all concentrations and decreased with increasing concentration of Yb(III) ions. The minimum Yb–Yb distance of the polymer of 10.4 Å suppresses long-range energy migration. For this reason, the concentration mechanism is in the limited-diffusion regime. Theoretical calculation of emission lifetimes was performed for the killer-site model and phonon-assisted energy transfer relaxation model. The calculated and experimental results were qualitatively more consistent in the latter model, suggesting that $[Yb_xGd_{1-x}(hfa)_3dpbp]_n$ acts as a high-purity host in terms of luminescence. This report is another step forward in understanding the fundamentals of lanthanide chemistry, especially lanthanide coordination polymers and lanthanide-doped inorganic hosts [44].

## 3.6 Appendix

### 3.6.1 Absorption Spectra of Yb(III) Ions

Diffuse reflectance (absorption spectra) of the Yb(III) ions in $[Yb_1(hfa)_3dpbp]_n$ and $[Yb_{0.01}Gd_{0.99}(hfa)_3dpbp]_n$ are provided in Fig. 3.11 to compare the spectral shape. The small absorption of the $[Yb_{0.01}Gd_{0.99}(hfa)_3dpbp]_n$ due to low concentration of Yb(III) ions rendered the comparison unreliable. However, the spectral shape seems to be identical.

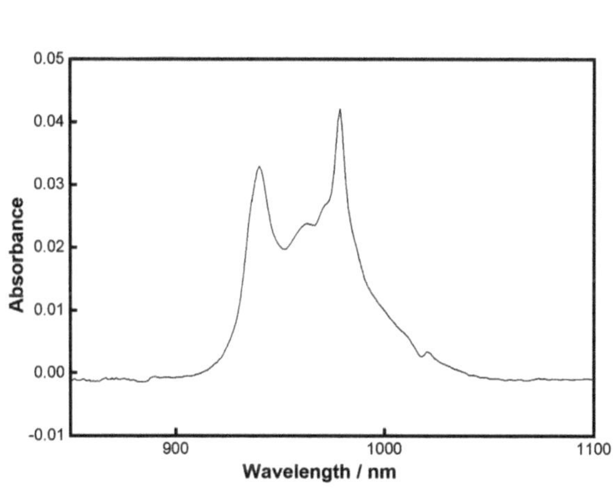


**Fig. 3.11** Absorption spectrum of $Yb(hfa)_3(TPPO)_2$ in $1.0 \times 10^{-2}$ M chloroform solution

### 3.6.2 *MATLAB Script for Calculation of Theoretical Lifetimes*

Calculations of theoretical lifetimes were performed using MATLAB software. The procedures are explained in Sect. 3.3.2. Below is an example of the script for the calculation of $[Yb_{0.5}Gd_{0.5}(hfa)_3dpbp]_n$ in the PAETR model. The section of the script defining the rate constants as well as the interaction matrix $\boldsymbol{A}$ is omitted since it is a tedious work that was done on a spreadsheet software.

```
clear

how_long = tic
clear CLEAN

=> Defining various rate constants here
=> Defining Interaction Matrix A here

m = 500                              % Cycle of calculation

%Yb0.5
S1 = sym(zeros(1, m))
clear S2 t
syms t S2(t)
S2(t) = 0

wb = waitbar(0, 'Calculating for x = 0.5')
CLEAN = onCleanup(@() close(wb))

for n = 1:m
   waitbar(n/m, wb, 'Calculating for x = 0.5');
   A2 = A1

% Random generation of a quenching center(s)
   h = 0                             % max# of QCs
   r = h + 1
   B1 = setdiff(1:135,68) .'
   q = randi(r,1) - 1                % Random number of QCs
   if q == 0
      B2 = []                        % Do nothing to B1
   else
      B2 = zeros(q,1)
      for i = 1:q
         k = randi(length(B1),1)
         B2(i) = B1(k)
         B1(k) = []
         A2(:, B2(i)) = 0
         A2(B2(i), :) = A2(B2(i), :)*1000    % Strong Quenching Center
         A2(B2(i),B2(i)) = -Anrad
      end
   end

% Inhomogeneous distribution of Yb ions
  g = 67 - q                 % 0(Yb1), 67(Yb0.5), 121(Yb0.1), 128(Yb0.05), 134(Yb0.01)
   if g <= 0
      for i = 1:length(B1)
```

```
            A2(B1(i), B1(i)) = 0
                    A2(B1(i), B1(i)) = -(Arad + Anrad + sum(A2(:, B1(i))) + (3.67*10^-
3)*sum(A2(:, B1(i))))
        end
    else
        B3 = zeros(g,1)
        for i = 1:g
            k = randi(length(B1),1)
            B3(i) = B1(k)
            B1(k) = []
            A2(:, B3(i)) = 0
            A2(B3(i), :) = 0
        end
        for i = 1:length(B1)
            A2(B1(i), B1(i)) = 0
                    A2(B1(i), B1(i)) = -(Arad + Anrad + sum(A2(:, B1(i))) + (3.67*10^-
3)*sum(A2(:, B1(i))))
        end
    end
        A2(68, 68) = 0
        A2(68, 68) = -(Arad + Anrad + sum(A2(:, 68)) + (3.67*10^-3)*sum(A2(:, 68)))

% Rate equation calculation
    I = eye(135)
    C = [0; 0; 0; 0; 0; 0; 0; 0; 0; 0; 0; 0; 0; 0; 0; 0; 0; 0; 0; 0; 0; 0; 0; 0; 0; 0; 0; 0; 0; 0; 0; 0; 0;
0; 0; 0; 0; 0; 0; 0; 0; 0; 0; 0; 0; 0; 0; 0; 0; 0; 0; 0; 0; 0; 0; 0; 0; 0; 0; 0; 0; 0; 0; 0; 0; 0; 0; 1; 0;
0; 0; 0; 0; 0; 0; 0; 0; 0; 0; 0; 0; 0; 0; 0; 0; 0; 0; 0; 0; 0; 0; 0; 0; 0; 0; 0; 0; 0; 0; 0; 0; 0; 0; 0; 0;
0; 0; 0; 0; 0; 0; 0; 0; 0; 0; 0; 0; 0; 0; 0; 0; 0; 0; 0; 0; 0; 0; 0; 0; 0; 0; 0; 0; 0; 0]
    [P,D] = eig(A2)
    V = (P*expm(D*t)*P^-1)*C
    S1(n) = sum(V(B1,1)) + V(68,1)
end                                     % End of entire cycle

% Storing to final rate equation
    S2(t) = sum(S1)

clear CLEAN

diary EEPCYb05.txt
S2(t)

% Graph
t = 0:0.0000001:0.00003
y = vpa(S2(t))
plot(t,y,'color','y','Linewidth',3)
Title = title('Yb Concentration Quenching','FontSize',15)
ylabel('Population Density','FontSize',10)
xlabel('Time / s','FontSize',10)
axis([0 0.00003 0 m])

S1

toc(how_long)

diary off
```

The calculation for each model took anywhere from 3–6 days to complete (on a 64-bit Windows 7, Intel Xeon E3-1241 v3 (8 M cache, 3.50 GHz, 4 cores, 8 threads), 32 GB RAM computer) depending on the concentration of Yb(III) ions, the number of erasing operations done to the interaction matrix *A* to account for the inhomogeneity, and the complexity of the equation for each cycle. For example, the calculation for the KS model took approximately 4.6 days to complete while the calculation for the PAETR model took approximately 6.1 days to complete. It should be noted that the codes were not output suppressed (using ";" at the end of each line) in order to keep track of the calculation. Using the output suppression may improve calculation speed.

### 3.6.3 Quantitative Considerations of Theoretical Lifetimes

The plot of the experimental (Fig. 3.7b) and theoretical lifetimes (Fig. 3.9) differ in the order of decrease; the former decreased from 21.7 to 19.2 μs while the latter decreased from 21.69 to only 21.65 μs. As explained in Sect. 3.4.3, this is likely due to assumed oscillator strength and Huang-Rhys parameter. In order to account for this, the absorption spectrum of $Yb(hfa)_3(TPPO)_2$ [45] was measured to estimate the oscillator strength. This Yb(III) complex have similar ligands (hfa and phosphine oxide ligands) that are coordinated in a similar geometrical form as that of $[Yb(hfa)_3dpbp]_n$. Thus, the oscillator strength between these two complexes are likely similar as well.

The absorption spectra of $Yb(hfa)_3(TPPO)_2$ in $1.0 \times 10^{-2}$ M chloroform solution ($n = 1.43$ at 1000 nm [46–48]) is shown in Fig. 3.11. The oscillator strength calculated from the area of the absorption spectra was $7.24 \times 10^{-6}$. This is more than twice as stronger than the assumed oscillator strength ($3 \times 10^{-6}$), which means the energy transfer rate constant is significantly increased and thus the PAETR rate constant as well.

The Huang-Rhys parameter was determined from absorption and emission spectra, but it is known from previous reports that this will not provide a very accurate value, and instead should be calculated from the ratio of intensity of phonon sideband and zero-phonon band [49–51]. Because of the complex absorption spectra (Fig. 3.10), it is difficult to consider the phonon sideband. As such, we instead assume a value. Huang-Rhys parameters can range from $10^{-2}$ to $10^{-1}$ order in lanthanides [49–51]. Here, the $s = 2 \times 10^{-1}$ were used based on ref [50]. Also, in order to keep the discussion general, the refractive index was set to 1. The reconsidered parameters are given in Table 3.2.

The recalculated theoretical lifetimes for PAETR model is shown in Fig. 3.12. The decay profile is still shown to be single exponential for all concentration and gradually decreases with increasing concentration. The lifetimes decreased from 21.69 to 18.31 μs. The decrease is now well within the order of decrease seen in the experiment.

**Table 3.2** Values reconsidered for the theoretical calculation

| Constant | Value | Constant | Value |
|---|---|---|---|
| $\chi_{abs}$ | 1 | $\hbar\omega$[a] | 3141 $cm^{-1}$ (ArC–H) |
| $f_{total}$ | $7.24 \times 10^{-6}$ ($n = 1.43$) | | 3068 $cm^{-1}$ (ArC–H) |
| $f_{MD}$ | $3.808 \times 10^{-7}$ | | 1654 $cm^{-1}$ (C=O) |
| $f_{ED}$ | $6.981 \times 10^{-6}$ | | 1599 $cm^{-1}$ (ArC=C) |
| $A_{tot}$ | $4.611 \times 10^{4}\ s^{-1}$ | | 1551 $cm^{-1}$ (ArC=C) |
| $\sum_{k=2,4,6} \Omega_k \left\|\langle l^N SLJ \\| U^{(k)} \\| l^N S'L'J' \rangle\right\|^2$ | $2.732 \times 10^{-20}\ cm^2$ | | 1527 $cm^{-1}$ (ArC=C) |
| $A_{rad(MD)}$ | $3.808 \times 10^{1}\ s^{-1}$ | | 1439 $cm^{-1}$ (P–C) |
| $A_{rad(ED)}$ | $1.028 \times 10^{2}\ s^{-1}$ | | 1254 $cm^{-1}$ (P=O) |
| $A_{nrad}$ | $4.504 \times 10^{4}\ s^{-1}$ | | 1200 $cm^{-1}$ (C–F) |
| $W_{ET(dd)}$ | $7.729 \times 10^{9} \times R^{-6}\ [s^{-1}]$ | $N$[a] | 3 (ArC–H) |
| $W_{ET(dq)}$ | $3.640 \times 10^{11} \times R^{-8}\ [s^{-1}]$ | | 3 (ArC–H) |
| $W_{ET(qq)}$ | $3.201 \times 10^{13} \times R^{-10}\ [s^{-1}]$ | | 6 (C=O) |
| $s$ | 0.2 | | 6 (ArC=C) |
| $W_{PAETR}$ | $W_{tot} \times 7.443 \times 10^{-2} [s^{-1}]$ | | 6 (ArC=C) |
| | | | 6 (ArC=C) |
| | | | 6 (P–C) |
| | | | 8 (P=O) |
| | | | 8 (C–F) |

[a]Vibration below ArC–H have minor influence on the value of $W_{PAETR}$

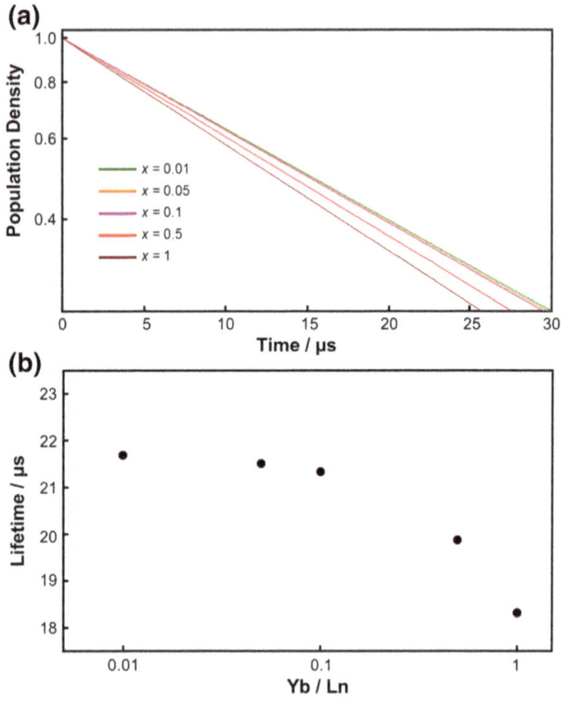


**Fig. 3.12** **a** Recalculated theoretical time-evolution of the population density of the excited state Yb(III) ions of $[Yb_xGd_{1-x}(hfa)_3dpbp]_n$ coordination polymers for $x = 0.01$–1 for PAETR model. **b** Plot of the theoretical lifetimes for the PAETR model

## References


1. Dexter DL, Schulman JH (1954) Theory of concentration quenching in inorganic phosphors. J Chem Phys 22:1063–1070
2. Di Bartolo B, Goldberg V (1980) Radiationless processes. Plenum Press
3. Auzel F, Baldacchini G, Laversenne L, Boulon G (2003) Radiation trapping and self-quenching analysis in $Yb^{3+}$, $Er^{3+}$, and $Ho^{3+}$ doped $Y_2O_3$. Opt Mater 24:103–109
4. Benz F, Strunk HP (2012) Rare earth luminescence: a way to overcome concentration quenching. AIP Adv 2:2–7
5. Wei W, Chen G, Baev A, He GS, Shao W, Damasco J, Prasad PN (2016) Alleviating luminescence concentration quenching in upconversion nanoparticles through organic dye sensitization. J Am Chem Soc 138:15130–15133
6. Yang P, Deng P, Yin Z (2002) Concentration quenching in Yb:YAG. J Lumin 97:51–54
7. Ito M, Boulon G, Bensalah A, Guyot Y, Goutaudier C, Sato H (2007) Spectroscopic properties, concentration quenching, and prediction of infrared laser emission of $Yb^{3+}$-doped $KY_3F_{10}$ cubic crystal. J Opt Soc Am B 24:3023–3032
8. Jacobsohn LG, Bennett BL, Muenchausen RE, Smith JF, Cooke DW (2007) Luminescent properties of nanophosphors. Radiat Meas 42:675–678
9. Cui Y, Chen B, Qian G (2014) Luminescent properties of nanophosphors. Coord Chem Rev 273–274:76–86
10. Heine J, Müller-Buschbaum K (2013) Engineering metal-based luminescence in coordination polymers and metal–organic frameworks. Chem Soc Rev 43:9232–9242
11. Yan B, Bai Y, Chen Z (2005) Synthesis, structure and luminescence of novel 1D chain coordination polymers $[Ln(isophth)(Hisophth)(H_2O)_4 \cdot 4H_2O]_n$ (Ln = Sm, Dy). J Mol Struct 741:141–147

12. Auzel F, Bonfigli F, Gagliari S, Baldacchini G (2001) The interplay of self-trapping and self-quenching for resonant transitions in solids; role of a cavity. J Lumin 94–95:293–297
13. Knoester J, Van Himbergen JE (1987) On the theory of concentration self-quenching by statistical traps. J Chem Phys 86:3571
14. Auzel F (2004) Upconversion and anti-stokes processes with f and d ions in solids. Chem Rev 104:139–173
15. Auzel F (2002) A fundamental self-generated quenching center for lanthanide-doped high-purity solids. J Lumin 100:125–130
16. Miyakawa T, Dexter DL (1970) Phonon sidebands, multiphonon relaxation of excited states, and phonon-assisted energy transfer between ions in solids. Phys Rev B 1:2961–2969
17. Balda R, Fernández J, Saéz de Ocáriz I, Adam JL, Mendioroz A, Montoya E (1999) Spectroscopic properties of $Pr^{3+}$ ions in fluorophosphate glass. Opt Mater 13:159–165
18. Utochnikova VV, Kalyakina AS, Bushmarinov IS, Vashchenko AA, Marciniak L, Kaczmarek AM, Van Deun R, Bräse S, Kuzmina NP (2016) Lanthanide 9-anthracenate: solution processable emitters for efficient purely NIR emitting host-free OLEDs. J Mater Chem C 4:9848–9855
19. Miyata K, Konno Y, Nakanishi T, Kobayashi A, Kato M, Fushimi K, Hasegawa Y (2013) Chameleon luminophore for sensing temperatures: control of metal-to-metal and energy back transfer in lanthanide coordination polymers. Angew Chem Int Ed 52:6413–6416
20. Hirai Y, Nakanishi T, Miyata K, Fushimi K, Hasegawa Y (2014) Thermo-sensitive luminescent materials composed of Tb (III) and Eu (III) complexes. Mater Lett 130:91–93
21. Miyata K, Ohba T, Kobayashi A, Kato M, Nakanishi T, Fushimi K, Hasegawa Y (2012) Thermostable organo-phosphor: low-vibrational Coordination polymers that exhibit different intermolecular interactions. ChemPlus Chem 77:277–280
22. Judd BR (1962) Optical absorption intensities of rare-earth ions. Phys Rev 127:750–761
23. Ofelt GS (1962) Intensities of crystal spectra of rare-earth ions. J Chem Phys 37:511–520
24. Hehlen MP, Brik MG, Krämer KW (2013) 50th anniversary of the Judd–Ofelt theory: an experimentalist's view of the formalism and its application. J Lumin 136:221–239
25. Kushida T (1973) Energy transfer and cooperative optical transitions in rare-earth doped inorganic materials. I. Transition probability calculation. J Phys Soc Jpn 34:1318–1326
26. Malta OL (2008) Mechanisms of non-radiative energy transfer involving lanthanide ions revisited. J Non Cryst Solids 354:4770–4776
27. Aboshyan-Sorgho L, Nozary H, Aebischer A, Bünzli JCG, Morgantini PY, Kittilstved KR, Hauser A, Eliseeva SV, Petoud S, Piguet C (2012) Optimizing millisecond time scale near-infrared emission in polynuclear chrome (III)–lanthanide (III) complexes. J Am Chem Soc 134:12675–12684
28. Starzak ME (1989) Mathematical methods in chemistry and physics. Springer
29. Bünzli JCG, Eliseeva SV (2011) Basics of lanthanide photophysics, in Springer series on fluorescence. Springer
30. Ito M, Goutaudier C, Guyot Y, Lebbou K, Fukuda T, Boulon G (2004) Crystal growth, $Yb^{3+}$ spectroscopy, concentration quenching analysis and potentiality of laser emission in $Ca_{1-X}Yb_XF_{2+X}$. J Phys: Condens Matter 16:1501–1521
31. Auzel F, de Sá GF, de Azevedo WM (1980) An example of concentration sensitive electron-phonon coupling in $\{(C_4H_9)_4N\}_3Eu_xY_{1-x}(NCS)_6$ and a new hypothesis for self-quenching. J Lumin 21:187–192
32. Auzel F, Pellé F (1996) Concentration and excitation effects in multiphonon non-radiative transitions of rare-earth ions. J Lumin 69:249–255
33. Vink AP, Meijerink A (1998) Electron–phonon coupling of $Cr^{3+}$-pairs and isolated sites in $\alpha$-$Al_2O_3$ and MgO. Spectrochim Acta, Part A 54:1755–1761
34. Carnall WT, Fields PR, Rajnak K (1965) Spectral intensities of trivalent lanthanides and actinides in solution I. $Pr^{3+}$, $Nd^{3+}$, $Er^{3+}$, $Tm^{3+}$, $Yb^{3+}$. 42:3797–3806
35. Auzel F, Jean-Louis AM, Toudic Y (1989) Oscillator strengths, quantum efficiencies, and laser cross sections of $Yb^{3+}$ and $Er^{3+}$ in III-V compounds. J Appl Phys 66:3952–3955
36. Krishnaiah KV, Rajeswari R, Kumar KU, Babu SS, Martin IR, Jayasankar CK (2014) Spectroscopy and radiation trapping of $Yb^{3+}$ ions in lead phosphate glasses. J Quant Spectrosc Radiat Transf 140:37–47

37. Seregina EA, Seregin AA, Tikhonov GV (2014) Spectral and luminescent characteristics of trivalent lanthanide ions in a $POCl_3$-$SnCl_4$ inorganic solvent. Opt Spectrosc 116:438–453
38. Balaji S, Mandal AK, Annapurna K (2012) Energy transfer based NIR to visible upconversion: Enhanced red luminescence from $Yb^{3+}$/$Ho^{3+}$ co-doped tellurite glass. Opt Mater 34:1930–1934
39. Lu W-G, Zhong D-C, Jiang L, Lu T-B (2012) Lanthanide coordination polymers constructed from Imidazole-4, 5-dicarboxylate and sulfate: syntheses, structural diversity, and photoluminescent properties. Cryst Growth Des 12:3675–3683
40. Decadt R, van Hecke K, Depla D, Leus K, Weinberger D, van Driessche I, Van Der Voort P, van Deun R (2012) Synthesis, crystal structures, and luminescence properties of carboxylate based rare-earth coordination polymers. Inorg Chem 51:11623–11634
41. Chen S, Fan R, Sun C, Wang P, Yang Y, Su Q, Mu Y (2012) Synthesis, structure, and luminescent properties of lanthanide-based two-dimensional and three-dimensional metal–organic frameworks with 2, 4′-biphenyldicarboxylic acid. Cryst Growth Des 12:1337–1346
42. Li ZY, Dai JW, Wang N, Qiu HH, Yue ST, Liu YL (2010) A series of three-dimensional 4d−4f heterometallic coordination polymers with six-connected doubly interpenetrated pcu net topology: structural, photoluminescent, and magnetic properties. Cryst Growth Des 10:2746–2751
43. Ramya AR, Sharma D, Natarajan S, Reddy MLP (2012) Highly luminescent and thermally stable lanthanide coordination polymers Designed from 4-(dipyridin-2-yl)aminobenzoate: efficient energy transfer from $Tb^{3+}$ to $Eu^{3+}$ in a mixed lanthanide coordination compound. Inorg Chem 51:8818–8826
44. Omagari S, Nakanishi T, Hirai Y, Kitagawa Y, Seki T, Fushimi K, Ito H, Hasegawa Y (2018) Origin of concentration quenching in ytterbium coordination polymers: phonon-assisted energy transfer. Eur J Inorg Chem, 561–567
45. Kishimoto S, Nakagawa T, Kawai T, Hasegawa Y (2011) Enhanced near-infrared luminescence of Yb (III) complexes with phosphine oxide and hexafluoroacetylacetonate ligands. Bull Chem Soc Jpn 84:1–5
46. Samoc A (2003) Dispersion of refractive properties of solvents: Chloroform, toluene, benzene, and carbon disulfide in ultraviolet, visible, and near-infrared. J Appl Phys 94:6167–6174
47. El-Kashef H (2002) Study of the refractive properties of laser dye solvents: toluene, carbon disulphide, chloroform, and benzene. Opt Mater 20:81–86
48. Kedenburg S, Vieweg M, Gissibl T, Giessen H (2012) Linear refractive index and absorption measurements of nonlinear optical liquids in the visible and near-infrared spectral region. Opt Mater Express 2:1588
49. de Mello Donegá C, Meijerink A, Blasse G (1992) Vibronic transition probabilities in the excitation spectra of the $Pr^{3+}$ ion. J Phys Condens Matter 4:8889–8902
50. de Mello Donegá C, Schenker S, Folkerts HF, Meijerink A, Blasse G (1994) Vibronic spectroscopy of $Pr^{3+}$ in host lattices with the scheelite structure. J Phys Condens Matter 6:6043–6056
51. Hehlen MP, Kuditcher A, Rand SC, Tischler MA (1997) Electron–phonon interactions in CsCdBr3:$Yb^{3+}$. J Chem Phys 107:4886–4892

# Chapter 4
# Spin-Orbit Coupling and Energy Transfer in Nonanuclear Lanthanide Clusters

**Abstract** Spin-orbit coupling is essential in lanthanide complexes for raising the yield of triplet excited state where the energy transfer to lanthanide ion mainly proceeds from, and therefore the sensitization efficiency of the lanthanide ion. However, since spin-orbit coupling mixes states of different multiplicity, strong spin-orbit coupling means that the decay of triplet excited state to the ground state is also enhanced, lowering the sensitization efficiency. This chapter explains the effect of spin-orbit coupling on the sensitization efficiency in $[Ln_9(\mu\text{-}OH)_{10}(\text{butyl salicylate})_{16}]NO_3$ ($Ln = Yb_nGd_{9-n}/Yb_nLu_{9-n}$, $n = 0, 1, 3, 7, 9$) where Gd(III) and Lu(III) ions tune the spin-orbit coupling strength. It is revealed that a stronger spin-orbit coupling does not necessarily lead to higher sensitization efficiency.

## 4.1 Introduction

Spin-orbit coupling (SOC) plays an important role in Ln(III) complexes to efficiently produce the $T_1$ state of organic ligands, which is the state that FET proceeds from [1–3]. SOC mixes states of different multiplicity, and thus $S_1$ and $T_1$, as well as $T_1$ and $S_0$ states are mixed (see Sect. 4.5.1) [4, 5]. This makes ISC and phosphorescence partially allowed processes. There are two ISC processes $S_1 \rightarrow T_1$ and $T_1 \rightarrow S_0$. So the relaxation of the $T_1$ state includes partially allowed phosphorescence and ISC to the ground state.

The SOC in Ln(III) complexes is strong due to (1) the large effective nuclear charge of the Ln(III) ions (termed as the "heavy atom effect"), and (2) the spin and orbital angular momentum of 4f-electrons (termed as the "paramagnetic effect"). As a result, $S_1 \rightarrow T_1$ ISC rate in Ln(III) complexes are assumed to be fast and ranges anywhere between $10^7$ to $10^{10}$ $s^{-1}$ [6–8]. S. Tobita and D. M. Guldi both showed in their respective reports that within the lanthanide series, the difference in the SOC in the complexes arises predominantly from the paramagnetic effect rather than the heavy atom effect. For example, Gd(III) complex with methyl salicylate ligands showed an order of magnitude larger $S_1 \rightarrow T_1$ ISC and $T_1 \rightarrow S_0$ relaxation rates compared to Lu(III) complex. This is in spite of the Lu(III) ion being the heavier atom. D. M. Guldi and his coworkers further demonstrated that both $S_1 \rightarrow T_1$ ISC

 
S. Omagari, *Energy Transfer Processes in Polynuclear Lanthanide Complexes*, Springer Theses, https://doi.org/10.1007/978-981-13-6049-7_4

and $T_1 \rightarrow S_0$ relaxation rates are roughly proportional to the magnetic moment of the lanthanide ions [8].

Here, the effect of SOC on the sensitization efficiency of Ln(III) complexes is considered. As mentioned in Sect. 1.3.1, the sensitization efficiency $\eta_{sens}$ is given by the following equation:

$$\eta_{sens} = \eta_{ISC} \times \eta_{FET} = \frac{k_{ISC}}{kr_{S_1} + knr_{S_1} + k_{ISC}} \times \frac{k_{FET}}{kr_{T_1} + knr_{T_1} + k_{FET}}, \tag{4.1}$$

where $k_{ISC}$, $k_{FET}$, are $S_1 \rightarrow T_1$ ISC, and FET rate constants, respectively. $kr$ and $knr$ are radiative and nonradiative rate constants, respectively, and the subscript denotes the states in which they apply to. Since SOC raises $k_{ISC}$, $kr_{T_1}$, and $knr_{T_1}$, it can be inferred from Eq. (4.1) that strong SOC enhances $\eta_{ISC}$ while reducing $\eta_{FET}$. The opposing effect that the SOC has on $\eta_{ISC}$ and $\eta_{FET}$ indicate that the sensitization efficiency $\eta_{sens}$, which is the product of $\eta_{ISC}$ and $\eta_{FET}$, should have an optimal point in terms of SOC strength.

Experimentally investigating the existence of the optimal point, however, is difficult. SOC is determined by the specific Ln(III) ion that is being used. Changing the Ln(III) ion also changes the FET rate constant $k_{FET}$ as well (all Ln(III) ions have different 4f-state energy levels) and thus, direct comparison of the effect of SOC induced by Ln(III) cannot be determined spectroscopically. Herein, nonanuclear Ln(III) clusters was selected as the suitable material [9]. The clusters allow control of SOC by mixing different Ln(III) ions without significantly distorting the structure. Yb(III), Gd(III), and Lu(III) ions were chosen for this study. The Yb(III) ion with its two-level 4f-states ($^2F_{7/2}$ and $^2F_{5/2}$ states) is suitable as the emissive center for the simplicity in discussing $T_1 \rightarrow Yb(^2F_{5/2})$ ET process (the FET). Gd(III) and Lu(III) ions were chosen as the non-emissive paramagnetic and diamagnetic centers, respectively.

In this chapter, the effect of SOC strength induced by paramagnetic and diamagnetic Ln(III) ions on the sensitization efficiency is described. Nonanuclear $Yb_nGd_{9-n}$ and $Yb_nLu_{9-n}$ clusters with butyl salicylate ligands ($[Ln_9(\mu\text{-}OH)_{10}$(butyl salicylate)$_{16}]NO_3$, Ln $= Yb_nGd_{9-n}/Yb_nLu_{9-n}$, $n = 0, 1, 3, 7, 9$) were synthesized. The clusters in this chapter are abbreviated as $Yb_nGd_{9-n}$-Bu and $Yb_nLu_{9-n}$-Bu (Bu = butyl salicylate ligands). The clusters were identified by IR, mass spectroscopy, and elemental analysis. The structural analysis was performed by the combination of powder and single-crystal XRD measurements. The effect of SOC was elucidated by emission spectra and lifetimes. This study explore whether the sensitization efficiency is improved or reduced in the presence of a strong SOC. Either result provides important insight into fundamental photophysics of Ln(III) complexes as well as factors to consider when designing them to optimize the energy transfer efficiency.

## 4.2 Experimental Section

### 4.2.1 *Material*

$Gd(NO_3)_3{\cdot}6H_2O$ (>99.95%) and $Lu(NO_3)_3{\cdot}4H_2O$ (>99.95) were purchased from Kanto Chemical Co., and $Yb(NO_3)_3{\cdot}5H_2O$ (99.9%) was purchased from Sigma-Aldrich Japan. Butylsalicylate was purchased from Tokyo Chemical Industries. Methanol for spectroscopy was purchased from Wako Pure Chemical Industries, Ltd. All reagents were used without further purification.

### 4.2.2 *Apparatus*

FAB-MS and ESI-MS spectra were measured on a JEOL JMS-700TZ and a JEOL JMS-T100LP, respectively. Elemental analyses were performed by an Exter Analytical CE440. Infrared spectra were recorded on a JASCO FT/IR-4600 spectrometer. XRD spectra were characterized by a RIGAKU SmartLab X-ray diffractometer. Single crystal X-ray diffractions were made on a RIGAKU RAXIS RAPID imaging plate area detector.

### 4.2.3 *Synthesis*

General synthetic procedures of $Yb_nGd_{9-n}$-Bu and $Yb_nLu_{9-n}$-Bu: The general synthetic procedures are shown in Fig. 4.1. 2.7 mmol of butyl salicylate ligand (Bu) was dissolved in methanol, and triethylamine (0.44 mL, 4.40 mmol) was added to this solution while stirred at 40 °C. 1.52 mmol of the appropriate ratio of $Gd(NO_3)_3{\cdot}6H_2O$ and $Yb(NO_3)_3{\cdot}5H_2O$ or $Lu(NO_3)_3{\cdot}4H_2O$ dissolved in methanol was added dropwise to this solution and further stirred for 60 min. The resulting white precipitate (the target cluster) was vacuum filtered and dried in vacuum. The clusters were recrystallized by evaporation method from a concentrated methanol solution.

**$Gd_9$-Bu**: Yield: 68%. Selected IR (ATR, $cm^{-1}$): 3573 (w, O–H), 3235 (w, O–H), 2956 (m, C–H), 1319 (s, C–O). FAB-MS: m/z = 4676.8 $[Gd_9(\mu\text{-}OH)_{10}(Bu)_{16}]^+$. Elemental analysis calculated for $C_{176}H_{218}NO_{61}Gd_9$: C, 44.61%, H, 4.64%, N, 0.30%. Found: C, 44.30%, H, 4.53%, N, <0.30%.
**$Yb_1Gd_8$-Bu**: Yield: 67%. Selected IR (ATR, $cm^{-1}$): 3572 (w, O–H), 3235 (w, O–H), 2957 (m, C–H), 1319 (s, C–O). FAB-MS: m/z = 4692.2 $[Yb_1Gd_8(\mu\text{-}OH)_{10}(Bu)_{16}]^+$. Elemental analysis calculated for $C_{176}H_{218}NO_{61}Yb_1Gd_8$: C, 44.46%, H, 4.62%, N, 0.29%. Found: C, 43.98%, H, 4.45%, N, <0.30%.
**$Yb_3Gd_6$-Bu**: Yield: 78%. Selected IR (ATR, $cm^{-1}$): 3572 (w, O–H), 3223 (w, O–H), 2957 (m, C–H), 1319 (s, C–O). FAB-MS: m/z = 4724.3 $[Yb_3Gd_6(\mu\text{-}OH)_{10}(Bu)_{16}]^+$.

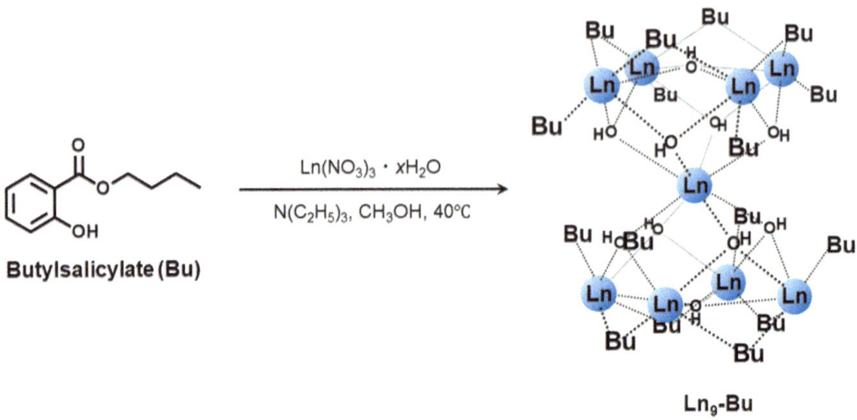


**Fig. 4.1** Synthetic procedure of Ln9-Bu. $Ln_9 = Yb_nGd_{9-n}$ or $Yb_nLu_{9-n}$

Elemental analysis calculated for $C_{176}H_{218}NO_{61}Yb_3Gd_6$: C, 44.17%, H, 4.59%, N, 0.29%. Found: C, 43.77%, H, 4.41%, N, <0.30%.

**$Yb_7Gd_2$-Bu:** Yield: 85%. Selected IR (ATR, $cm^{-1}$): 3582 (w, O–H), 3210 (w, O–H), 2957 (m, C–H), 1321 (s, C–O). FAB-MS: m/z = 4785.8 $[Yb_7Gd_2(\mu\text{-}OH)_{10}(Bu)_{16}]^+$. Elemental analysis calculated for $C_{176}H_{218}NO_{61}Yb_7Gd_2$: C, 43.59%, H, 4.53%, N, 0.29%. Found: C, 43.27%, H, 4.36%, N, <0.30%.

**$Yb_9$-Bu**: Yield: 78%. Selected IR (ATR, $cm^{-1}$): 3585 (w, O–H), 3206 (w, O–H), 2956 (m, C–H), 1320 (s, C–O). FAB-MS: m/z = 4819.0 $[Yb_9(\mu\text{-}OH)_{10}(Bu)_{16}]^+$. Elemental analysis calculated for $C_{176}H_{218}NO_{61}Yb_9$: C, 43.31%, H, 4.50%, N, 0.29%. Found: C, 42.99%, H, 4.37%, N, <0.30%.

**$Yb_7Lu_2$-Bu**: Yield: 80%. Selected IR (ATR, $cm^{-1}$): 3585 (w, O–H), 3226 (w, O–H), 2956 (m, C–H), 1321 (s, C–O). ESI-MS: m/z = 4822.74 $[Yb_7Lu_2(\mu\text{-}OH)_{10}(Bu)_{16}]^+$. Elemental analysis calculated for $C_{176}H_{218}NO_{61}Yb_7Lu_2$: C, 43.27%, H, 4.50%, N, 0.29%. Found: C, 42.94%, H, 4.37%, N, <0.30%.

**$Yb_3Lu_6$-Bu**: Yield: 28%. Selected IR (ATR, $cm^{-1}$): 3584 (w, O–H), 3206 (w, O–H), 2957 (m, C–H), 1321 (s, C–O). ESI-MS: m/z = 4830.73 $[Yb_3Lu_6(\mu\text{-}OH)_{10}(Bu)_{16}]^+$. Elemental analysis calculated for $C_{176}H_{218}NO_{61}Yb_3Lu_6$: C, 43.21%, H, 4.49%, N, 0.29%. Found: C, 42.73%, H, 4.31%, N, 0.25%.

**$Yb_1Lu_8$-Bu**: Yield: 81%. Selected IR (ATR, $cm^{-1}$): 3586 (w, O–H), 3208 (w, O–H), 2956 (m, C–H), 1321 (s, C–O). ESI-MS: m/z = 4834.75 $[Yb_1Lu_8(\mu\text{-}OH)_{10}(Bu)_{16}]^+$. Elemental analysis calculated for $C_{176}H_{218}NO_{61}Yb_1Lu_8$: C, 43.17%, H, 4.49%, N, 0.29%. Found: C, 43.04%, H, 4.36%, N, <0.30%.

**$Lu_9$-Bu**: Yield: 80%. Selected IR (ATR, $cm^{-1}$): 3586 (w, O–H), 3207 (w, O–H), 2957 (m, C–H), 1321 (s, C–O). FAB-MS: m/z = 4836.1 $[Lu_9(\mu\text{-}OH)_{10}(Bu)_{16}]^+$. Elemental analysis calculated for $C_{176}H_{218}NO_{61}Lu_9$: C, 43.16%, H, 4.49%, N, 0.29%. Found: C, 42.82%, H, 4.33%, N, <0.30%.

### 4.2.4 *Spectroscopy*

Absorption spectra of the clusters were obtained by using a JASCO V-670 spectrometer. Emission spectra and lifetimes were measured using a Horiba/Jobin-Yvon FluoroLog-3 ps spectrofluorometer. The emission spectra of UV-vis and NIR region were connected using $Eu(hfa)_3(TPPO)_2$ complex as a reference to match the detection level of both detectors. A combination of JASCO FP-6600 spectrometer and Oxford Instruments OptistatDN2 cryostat was used to measure emission spectra at low temperature. The absolute quantum yield was measured by a combination of JASCO FP-6600 and an integration sphere.

### 4.2.5 *Crystallography*

Colorless single crystal of $Gd_9$-Bu, $Yb_9$-Bu, and $Lu_9$-Bu obtained from solutions in methanol were mounted on a glass fiber by using epoxy resin glue. All measurements were made using a Rigaku RAXIS RAPID imaging plate area detector with graphite-monochromated MoK$\alpha$ radiation. Corrections for decay and Lorentz-polarization effects were made using a spherical absorption correction, solved by direct methods, and expanded using Fourier techniques. Non-hydrogen atoms were refined anisotropically except for disordered atoms. Hydrogen atoms were refined using the riding model. The final cycle of full-matrix least-squares refinement was based on observed reflections and variable parameters. All calculations were performed using a Crystal-Structure crystallographic software package. The CIF data were confirmed by using the checkCIF/PLATON service. CCDC-1479981 ("$Gd_9$ cluster"), CCDC-1578015 ("$Yb_9$ cluster"), and CCDC-1578016 ("$Lu_9$ cluster") contain the crystallographic data for this paper. These data can be obtained free of charge from The Cambridge Crystallographic Data Center via www.ccdc.cam.ac.uk/data_request/cif.

## 4.3 Results and Discussion

### 4.3.1 *Structure and Identification*

The crystal structures of $Gd_9$-Bu, $Yb_9$-Bu, and $Lu_9$-Bu are shown in Fig. 4.2 and the crystallographic data in Table 4.1. The crystallographic analysis showed that the Ln(III) ions in the cluster form an "hour-glass" structure in which the upper four and lower four Ln(III) ions are connected to the center Ln(III) ion via oxygen. All Ln (III) ions have an 8-coordination structure. The center Ln(III) ion is coordinated only by oxygen atoms and not by Bu ligands while each of the Ln(III) ions in the outer unit is coordinated by both Bu ligands and oxygen atoms. In the outer unit, the

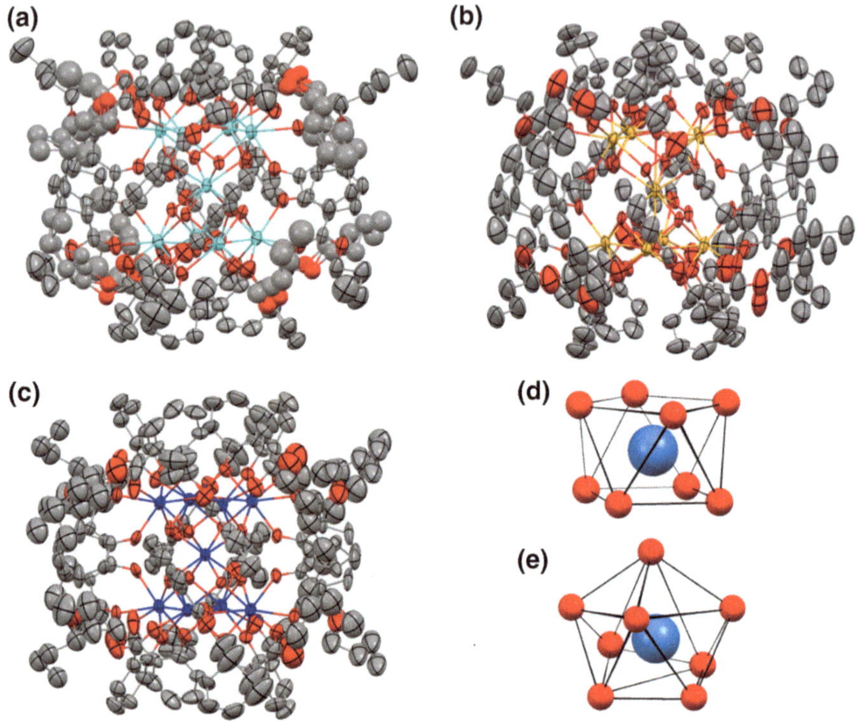


**Fig. 4.2** ORTEP drawing (50% probability displacement ellipsoids) of **a** $Gd_9$-Bu, **b** $Yb_9$-Bu, and **c** $Lu_9$-Bu. **d** center Ln(III) ion and **e** other Ln(III) ions (taken from $Gd_9$-Bu as an example)

upper and the lower parts are twisted approximately 45° with respect to the vertical axis that goes through the two $\mu_4$-O and the center Ln(III) ion.

Continuous shape measure analysis (CShM) was employed for the Ln(III) ions in the clusters to investigate the coordination geometry [10, 11]. Continuous Shape Measures (CShM) was calculated using SHAPE program [12]. The CShM criterion $S$ is given by the following equation:

$$S_{\mathrm{CShM}} = \min \frac{\sum_k^N \left| \boldsymbol{Q}_k - \boldsymbol{P}_k \right|^2}{\sum_k^N \left| \boldsymbol{Q}_k - \boldsymbol{Q}_0 \right|^2} \times 100, \tag{4.2}$$

where $\boldsymbol{Q}_k$ is the vertices of an actual structure, $\boldsymbol{Q}_0$ is the center of mass of an actual structure, $\boldsymbol{P}_k$ is the vertices of an ideal structure, and $N$ is the number of vertices. The $S$ value represents the degree of the deviation from a specified ideal coordination geometry. The smaller the value of $S$, the closer the structure is to a specified ideal geometry. For example, $S = 0$ means that the structure is fully identical with the ideal geometry. CShM were chosen over another method called "Shape Measure" (ShM) [13] because it accounts for the distortion of the center metal

**Table 4.1** Crystallographic information of $Gd_9$-Bu, $Yb_9$-Bu, and $Lu_9$-Bu

| | $Gd_9$-Bu | $Yb_9$-Bu | $Lu_9$-Bu |
|---|---|---|---|
| Chemical formula | $C_{176}H_{218}Gd_9NO_{61}$ | $C_{176}H_{208}O_{58}Yb_9$ | $C_{176}H_{208}O_{58}Lu_9$ |
| Formula weight | 4738.88 | 4808.90 | 4826.24 |
| Crystal system | Orthorhombic | Tetragonal | Tetragonal |
| Space group | *Pbcn* | $P4_2/n$ (#86) | $P4_2/n$ (#86) |
| $a$/Å | 29.1572(7) | 23.5449(3) | 23.857(3) |
| $b$/Å | 21.2906(5) | 23.5449(3) | 23.857(3) |
| $c$/Å | 31.7940(7) | 36.1157(7) | 36.2570(10) |
| $\alpha$/deg | 90 | 90 | 90 |
| $\beta$/deg | 90 | 90 | 90 |
| $\gamma$/deg | 90 | 90 | 90 |
| $V$/Å$^3$ | 19,736.9(8) | 20,021.2(5) | 20,636(4) |
| $Z$ | 4 | 4 | 4 |
| $D_{calcd}$/g cm$^{-3}$ | 1.595 | 1.595 | 1.553 |
| $T$/K | 123.15 | 93.15 | 173.15 |
| $\mu$ ($Mo_{K\alpha}$)/cm$^{-1}$ | 30.655 | 42.345 | 43.357 |
| Max 2θ/deg | 55.0 | 56.6 | 50.7 |
| No. of measured reflections | 183,154 | 74,859 | 120,501 |
| No. of unique reflections | 22,608 | 22,456 | 18,839 |
| $R^a$ | 0.1215 | 0.0696 | 0.1265 |
| $wR_2^b$ | 0.2134 | 0.1726 | 0.2115 |

$^a R = \Sigma||F_0|-|F_c||/\Sigma|F_0|$

$^b wR_2 = \left[\Sigma w\left(F_0^2-F_c^2\right)^2/\Sigma w\left(F_0^2\right)^2\right]^{1/2}$

ion from the center of mass (see Sect. 4.5.2). The $S$ values calculated for the clusters are summarized in Table 4.2. For the center Ln(III) ion (Fig. 4.2d), the $S$ values ranged from 0.077 to 0.105 when calculated for 8-coordinated square antiprism (8-SAP) geometry and 2.371 to 2.481 for 8-coordinated trigonal dodecahedron (8-TDH) geometry. Therefore, the coordination geometry of the center Ln(III) ions is attributed to the 8-SAP structure. There was no specific dependence on the ionic radii of Ln(III) ions on the $S$ values calculated for the clusters as 8-SAP. On the other hand, the other Ln(III) ions (Fig. 4.2e) showed smaller S values when calculated for 8-TDH than 8-SAP. Therefore, the coordination geometry of the Ln(III) ions other than the center Ln(III) is attributed to the 8-TDH structure. Unlike the center Ln(III) ion calculated for 8-SAP, the other Ln(III) ions showed larger S values for Ln(III) ions with larger ionic radii when calculated for 8-TDH. However, the differences are still quite small, and these clusters can be considered to have the same structure. These structural results are similar to that of the previously reported $Tb_9$-Bu [9].

**Table 4.2** Continuous Shape Measure calculation results for $Gd_9$-Bu, $Yb_9$-Bu, and $Lu_9$-Bu. Reprinted from Ref. [14] with permission from Elsevier

| | $Gd_9$ cluster | | $Yb_9$ cluster | | $Lu_9$ cluster | |
|---|---|---|---|---|---|---|
| | 8-TDH | 8-SAP | 8-TDH | 8-SAP | 8-TDH | 8-SAP |
| Ln1 | 2.976 | 4.725 | 2.573 | 4.578 | 2.565 | 4.455 |
| Ln2 | 2.836 | 4.813 | 2.545 | 4.546 | 2.481 | 4.633 |
| Ln3 | 2.704 | 4.588 | 2.413 | 4.776 | 2.447 | 4.622 |
| Ln4 | 2.513 | 4.822 | 2.404 | 4.623 | 2.400 | 4.507 |
| Ln5 | 2.976 | 4.725 | 2.573 | 4.578 | 2.565 | 4.455 |
| Ln6 | 2.836 | 4.813 | 2.545 | 4.546 | 2.481 | 4.633 |
| Ln7 | 2.704 | 4.588 | 2.413 | 4.776 | 2.447 | 4.622 |
| Ln8 | 2.513 | 4.822 | 2.404 | 4.623 | 2.400 | 4.507 |
| Ln9[a] | 2.481 | 0.082 | 2.396 | 0.077 | 2.371 | 0.105 |

[a]Center Ln(III) ion

As a crystal, these three clusters have different crystal system with $Yb_9$-Bu and $Lu_9$-Bu being triclinic while $Gd_9$-Bu being orthorhombic (Table 4.1). This may be due to the larger ionic radii of Gd(III) ions with minor influence on the microscopic level but is large on the larger scale. To compare the structure of the other clusters ($Yb_nGd_{9-n}$-Bu and $Yb_nLu_{9-n}$-Bu, n = 1, 3, and 7), XRD measurements were performed. The XRD patterns for the clusters shown in Fig. 4.3 are similar, but the peaks shift to the lower angle with increasing number of Yb(III) and Lu(III) ions compared to $Gd_9$-Bu, and the peak heights are quite different among the clusters. The spectroscopic measurements were performed in methanol solution, and intermolecular interaction is minimized. Therefore, any spectroscopic differences of the clusters should arise from the structure of a single molecule. This will become important for the later discussion because a distortion of ligand alignment may affect both the ligand and Yb(III) emissions.

### 4.3.2 *Photophysical Properties*

The emission intensity of the fluorescence and phosphorescence from the ligands in $Yb_nGd_{9-n}$-Bu and $Yb_nLu_{9-n}$-Bu provides information on photophysical properties of the ligands (i.e., SOC) under the influence of the paramagnetic effect of Ln(III) ions. On the other hand, the emission intensity of Yb(III) emission implies the sensitization efficiency $\eta_{sens}$. The emission spectra of $Yb_nGd_{9-n}$-Bu and $Yb_nLu_{9-n}$-Bu in $1.0 \times 10^{-4}$ M degassed methanol solution in the visible and NIR region excited at 371 nm are shown in Fig. 4.4a. The peak at 410 nm corresponds to the fluorescence from the butyl salicylate ligands. The fluorescence intensity is the lowest for $Yb_9$-Bu and increases with increasing number of Gd(III) or Lu(III) ions. Increasing

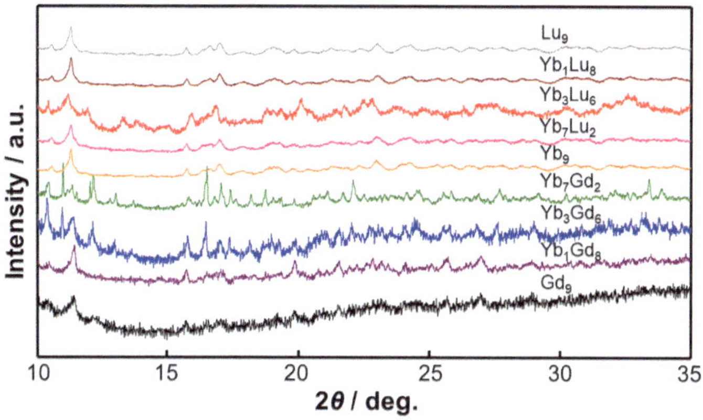


**Fig. 4.3** Powder XRD patterns of $Yb_nGd_{9-n}$-Bu and $Yb_nLu_{9-n}$-Bu

number of Lu(III) ions showed greater enhancement in emission intensity compared to those with Gd(III) ions (Fig. 4.4b). From these spectra, the strength of SOC in this cluster is in the order of Yb(III) > Gd(III) > Lu(III). A similar trend was reported by D. M. Guldi [8]. The absolute quantum yield of the fluorescence of $Lu_9$-Bu was below 1% (calculated as 0.2%). Phosphorescence was observed at low temperature (180°K) for $Gd_9$-Bu and $Yb_9$-Bu at around 465 nm (Fig. 4.5). The intensity of the phosphorescence in relative to the fluorescence is significantly reduced for $Yb_9$-Bu due to the presence of $T_1 \rightarrow$ Ln energy transfer. Phosphorescence was not observed for $Lu_9$-Bu because of the high fluorescence intensity, and low $S_1 \rightarrow T_1$ ISC and $T_1 \rightarrow S_0$ relaxation rate compared to those of $Gd_9$-Bu.

Emission peak at 976 nm in Fig. 4.4b originates from the ${}^2F_{5/2} \rightarrow {}^2F_{7/2}$ transition of Yb(III) ions. The spectral shape of the Yb(III) near-infrared (NIR) emission was identical for all compositions. The spectral shape of Ln(III) emission is known to be sensitive to the coordination geometry, thus the NIR emission spectra show that the coordination geometries around Yb(III) ions are identical for all clusters [15, 16]. The continuous shape measure calculations (see Sect. 4.5.2) support this analysis. Therefore, the radiative rate constants of Yb(III) ions in these clusters are equal.

The NIR emission lifetimes monitored at 976 nm ($\lambda_{EX} = 371$ nm) were single-exponential for all compositions (Fig. 4.6a). Emission lifetime values derived from single-exponential fitting showed that the lifetimes of the clusters gradually increased with decreasing number of Gd(III) ions and increasing number of Lu(III) ions (Fig. 4.6b). In particular, the emission lifetimes of $Yb_1Gd_8$-Bu, $Yb_9$-Bu, and $Yb_1Lu_8$-Bu were 568, 598, and 740 ns, respectively. The difference in the ionic radii of Gd(III), Yb(III), and Lu(III) ions may have contributed to the difference in the values of non-radiative rate constants of Yb(III) ions in the clusters.

Figure 4.7a shows the plot of the NIR emission intensity of the clusters, normalized by the intensity of $Yb_9$-Bu. The emission intensities in the plot indirectly mean the sensitization efficiency. The emission intensity of $Yb_1Gd_8$-Bu was 1.6 times

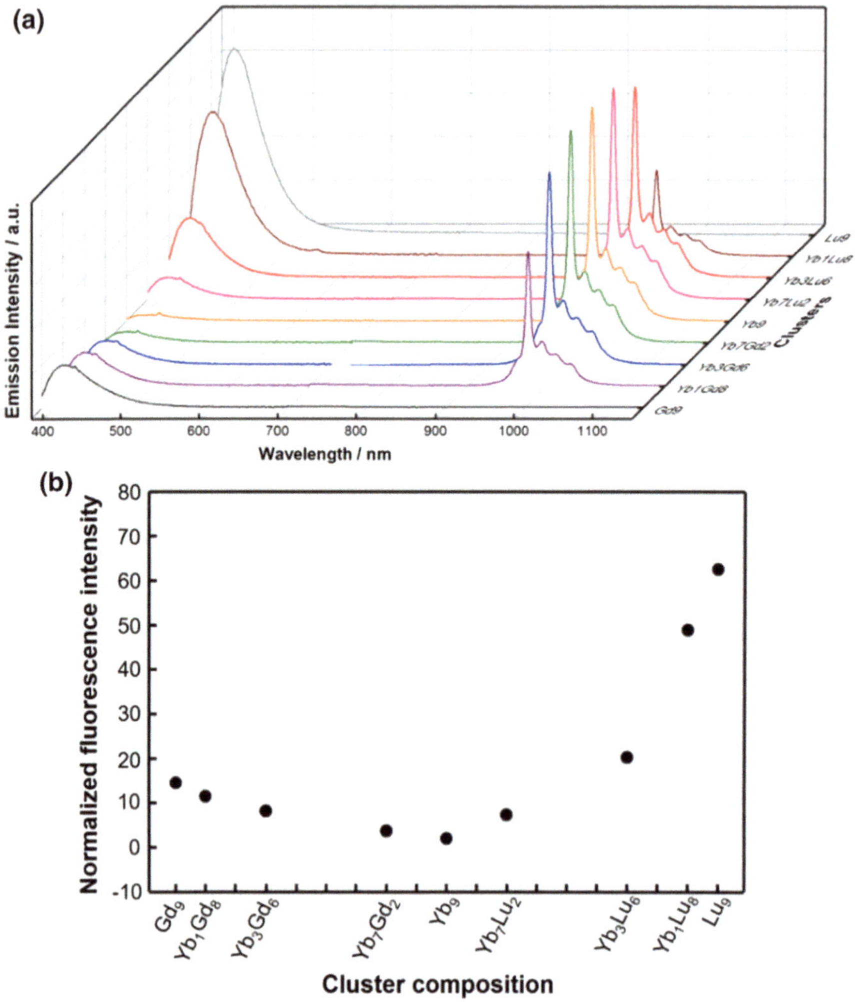


**Fig. 4.4** **a** Emission spectra of $Yb_nGd_{9-n}$-Bu/$Yb_nLu_{9-n}$-Bu in $1.0 \times 10^{-4}$ M methanol solution ($\lambda_{EX} = 371$ nm). The spectra for the UV-Vis and the NIR regions were measured separately, and connected by correcting the detectors with the previously reported $Eu(hfa)_3(TPPO)_2$ complex. **b** Plot of the emission intensity ($\lambda_{EM} = 410$ nm) of the ligand fluorescence. Reprinted from Ref. [14] with permission from Elsevier

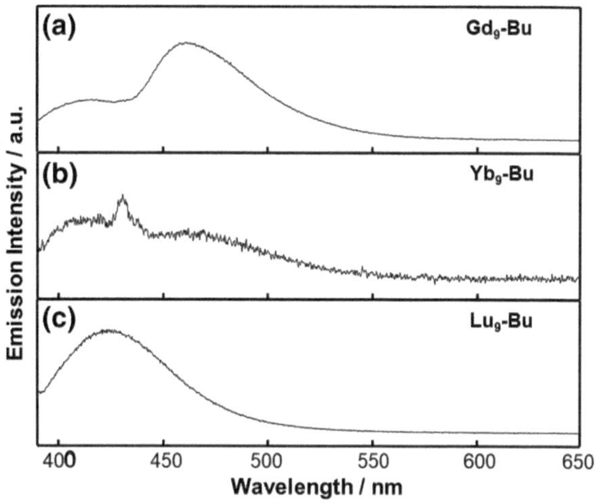


**Fig. 4.5** Emission spectra of **a** $Gd_9$-Bu, **b** $Yb_9$-Bu, and **c** $Lu_9$-Bu in $1.0 \times 10^{-4}$ M methanol solution at 210 K ($\lambda_{EX} = 371$ nm). A Raman peak is observed at around 430 nm for $Yb_9$-Bu

higher than that of $Yb_1Lu_8$-Bu. For these two clusters, the main contributor to the sensitization efficiency is the $S_1 \rightarrow T_1$ ISC efficiency due to the SOC induced by Gd(III) or Lu(III) ions (region A). Stronger SOC by Gd(III) ions leads to more efficient $S_1 \rightarrow T_1$ ISC (Fig. 4.7b, green arrow) in $Yb_1Gd_8$-Bu compared to that of $Yb_1Lu_8$-Bu. On the other hand, $Yb_7Gd_2$-Bu, $Yb_9$-Bu, and $Yb_7Lu_2$-Bu showed the same emission intensity. Normally, reduced number of emissive Ln(III) ions leads to a decrease in $T_1 \rightarrow$ Ln energy transfer efficiencies, but this phenomenon was not observed. Thus, for these three clusters, the $T_1 \rightarrow S_0$ relaxation is the main contributor to the sensitization efficiency (region B). Gd(III) and Lu(III) ions mitigated the fast $T_1 \rightarrow S_0$ relaxation induced by the Yb(III) ions (Fig. 4.7c, blue arrow) and thus the sensitization efficiency did not drop. These results suggest that there is an optimal strength of SOC for a high sensitization efficiency in Ln(III) complexes.

## 4.4 Conclusion

In this chapter, the effect of SOC strength induced by paramagnetic and diamagnetic Ln(III) ions on the sensitization efficiency in nonanuclear Ln(III) clusters was explored. Nonanuclear $Yb_nGd_{9-n}$-Bu and $Yb_nLu_{9-n}$-Bu ($[Ln_9(\mu\text{-}OH)_{10}(Bu)_{16}]^+NO_3^-$, Ln = $Yb_nGd_{9-n}/Yb_nLu_{9-n}$, $n = 0, 1, 3, 7, 9$) were synthesized and investigated their photophysical properties by emission spectra and lifetimes. The result implies that SOC affects sensitization efficiency both positively and negatively depending on the strength of SOC, and thus there is an optimal region in the SOC strength. It is also suggested that mixing Ln(III) ions with different SOC in Ln(III) clusters is an effective method to optimize sensitization efficiency. The findings are another step forward in understanding the fundamental photophysical

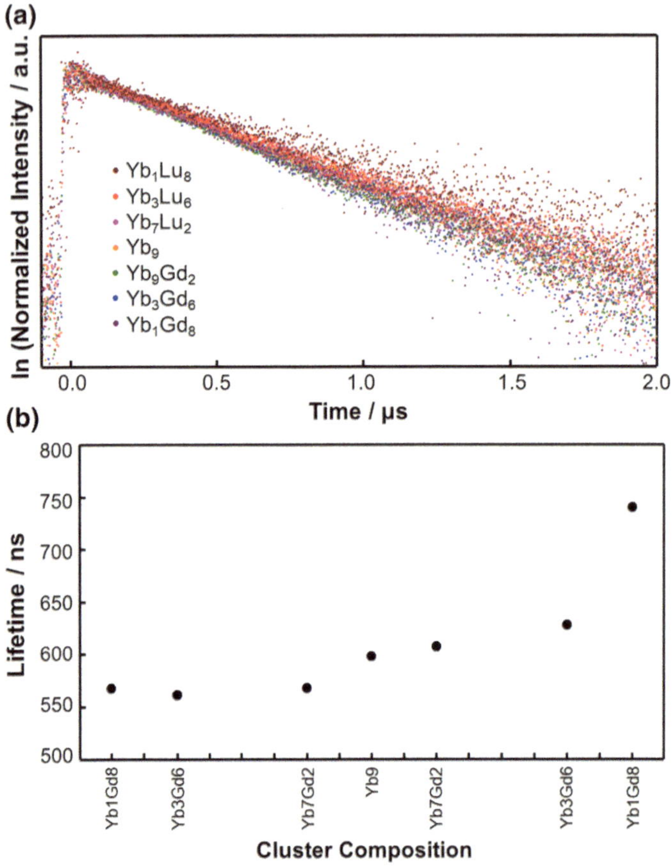


**Fig. 4.6** **a** Emission decay profiles of $Yb_nGd_{9-n}$-Bu/$Yb_nLu_{9-n}$-Bu in $1.0 \times 10^{-4}$ M methanol solution ($\lambda_{EX} = 371$ nm, $\lambda_{EM} = 976$ nm). **b** Lifetime plot of the clusters. Reprinted from Ref. [14] with permission from Elsevier

processes in Ln(III) complexes, as well as providing a possible novel strategy to enhance the luminescence efficiency [14].

## 4.5 Appendix

### *4.5.1 Notes on Spin-Orbit Coupling*

To be precise, spin-orbit coupling Hamiltonian $H_{SO}$ does not always mix $S_1$ and $T_1$ state since this depends on the symmetry of the two states. In fact, spin-orbit coupling may mix higher singlet excited states $S_n$ with another higher triplet excited states $T_n$.

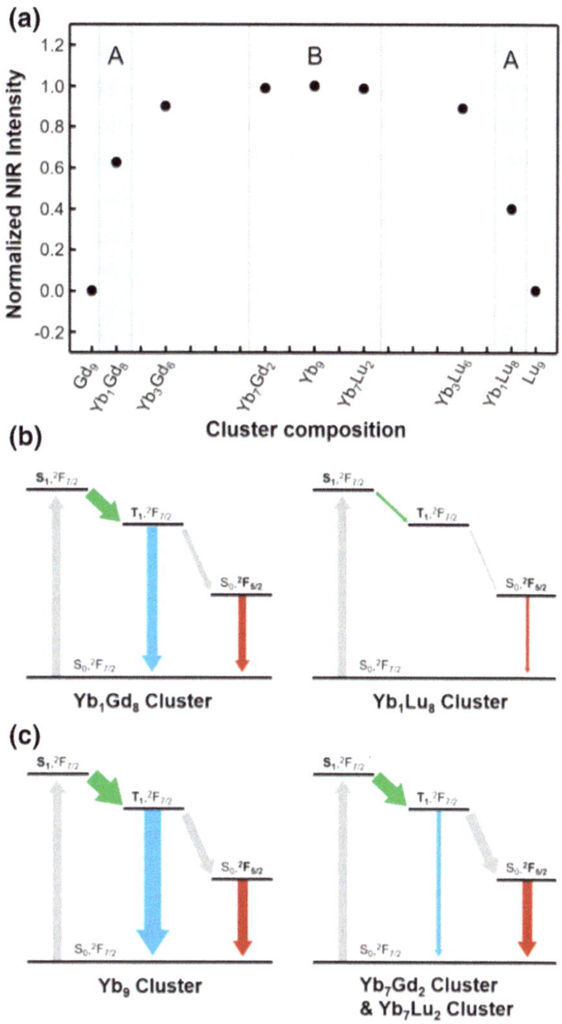


**Fig. 4.7** **a** Plot of the NIR emission intensities normalized by the peak intensity of $Yb_9$-Bu. The shaded area represents the two different regions in which the dominating photophysical process differs. **b** Comparison of the photophysical processes in $Yb_1Gd_8$-Bu and $Yb_1Lu_8$-Bu (region A) and **c** $Yb_9$-Bu, $Yb_7Lu_2$-Bu, $Yb_7Gd_2$-Bu (region B). The thickness of the arrows represents the yield resulting from the process (not the rate constant of the process). Reprinted from Ref. [14] with permission from Elsevier

For example, mixing between singlet and triplet $n\pi^*$ state as well as singlet and triplet $\pi\pi^*$ states is zero to the first order. On the other hand, mixing of $n\pi^*$ and $\pi\pi^*$ states of different multiplicity is large (known as the El-Sayed rule) [17]. Nonetheless, the convention is that the intersystem crossing is written as a process that occurs between $S_1$ to $T_1$ for simplicity [18]. Since this study focuses on nonanuclear $Yb_nGd_{9-n}$ and $Yb_nLu_{9-n}$ with the same butylsalicylate organic ligands (Bu), the mixing problem of $S_1$ and $T_1$ states can be neglected as long as the comparison is kept exclusively between these clusters.

### 4.5.2 Continuous Shape Measure and Shape Measure

Continuous shape measure (CShM) calculates the deviation of the vertices of an actual structure from the vertices of an ideal structure [10]. The CShM criterion $S_{\mathrm{CShM}}$ is given by the following equation:

$$S_{\mathrm{CShM}} = \min \frac{\sum_k^N |\boldsymbol{Q}_k - \boldsymbol{P}_k|^2}{\sum_k^N |\boldsymbol{Q}_k - \boldsymbol{Q}_0|^2} \times 100, \tag{4.2}$$

where $\boldsymbol{Q}_k$ is the vertices of an actual structure, $\boldsymbol{Q}_0$ is the center of mass of an actual structure, $\boldsymbol{P}_k$ is the vertices of an ideal structure, and $N$ is the number of vertices. In the case of $Gd_9$-Bu, for each Gd(III) ion, $N$ equals 9 (Gd(III) ion and eight oxygen atoms). CShM calculation takes accounts of the deviation of the metal atom from the center of mass.

On the other hand, shape measure (ShM) [13] calculates the deviation of the dihedral angles of an actual structure from the dihedral angles of an ideal structure. The ShM criterion $S_{\mathrm{ShM}}$ is given by the following equation:

$$S_{\mathrm{ShM}} = \min \sqrt{\frac{1}{M} \sum_i^M (\delta_i - \theta_i)^2} \times 100, \tag{4.3}$$

where $\delta_i$ is the angle of two planes of the ideal structure, $\theta_i$ is the angle of two planes of the actual structure, and $M$ is the number of dihedral angles. In the case of $Gd_9$-Bu, for each Gd(III) ion, $M$ equals 18. Since dihedral angles are independent of the effect of the distortion of the metal ion, ShM calculations do not take into consideration of the distortion of the metal ion.

Table 4.3 summarizes the difference in $S_{\mathrm{CShM}}$ values when the metal ion is taken into consideration to those of when the metal ion is not considered. The importance of considering the deviation of the metal atom from the center of mass of a structure lies in the fact that without such consideration, the $S$ values were smaller. This indicates that when the metal ion is not considered, the actual structure is calculated to be closer to ideal. Therefore, the inclusion of the metal ion is indeed important to derive the more accurate result. It is essential to point out that the $S_{\mathrm{CShM}}$ value of $Gd_9$ (center Gd(III) ion) calculated with and without the metal ion was almost identical. This implies that the distortion of the structure of $Gd_9$ was purely due to the position of the oxygen atoms.

**Table 4.3** Calculated $S_{CShM}$ values for all Gd(III) ions with and without including metal ion

| Gd(III) ion | $S_{SAP}$ (without) | $S_{SAP}$ (with) | $S_{TDH}$ (without) | $S_{TDH}$ (with) |
|---|---|---|---|---|
| Gd1 | 4.670 | 4.725 | 2.920 | 2.976 |
| Gd2 | 4.752 | 4.813 | 2.774 | 2.836 |
| Gd3 | 4.532 | 4.588 | 2.646 | 2.704 |
| Gd4 | 4.746 | 4.822 | 2.435 | 2.513 |
| Gd5 | 4.670 | 4.725 | 2.920 | 2.976 |
| Gd6 | 4.752 | 4.813 | 2.774 | 2.836 |
| Gd7 | 4.532 | 4.588 | 2.646 | 2.704 |
| Gd8 | 4.746 | 4.822 | 2.435 | 2.513 |
| Gd9 | 0.082 | 0.082 | 2.481 | 2.481 |

$S_{SAP}$ and $S_{TDH}$ are $S$ values calculated for an 8-coordinated square antiprism and an 8-coordinated trigonal dodecahedron, respectively. Label in parentheses indicates whether metal ion has been taken into consideration or not

# References


1. Moore EG, Samuel APS, Raymond KN (2009) From antenna to assay: lessons learned in lanthanide luminescence. Acc Chem Res 42:542–552
2. Huang, C (2010) Rare earth coordination chemistry: fundamentals and applications. Wiley
3. Lima NBD, Gonçalves SMC, Júnior SA, Simas AM (2013) A comprehensive strategy to boost the quantum yield of luminescence of europium complexes. Sci Rep 3:2395
4. Marian CM (2012) Spin–orbit coupling and intersystem crossing in molecules. WIREs Comput Mol Sci 2:187–203
5. Forster LS (2006) Intersystem crossing in transition metal complexes. Coord Chem Rev 250:2023–2033
6. Tobita S, Arakawa M, Tanaka I (1985) The paramagnetic metal effect on the ligand localized S1. apprx. fwdarw. T1 intersystem crossing in the rare-earth-metal complexes with methyl salicylate. J Phys Chem 89:5649–5654
7. Tobita S, Arakawa M, Tanaka I (1984) Electronic relaxation processes of rare earth chelates of benzoyltrifluoroacetone. J Phys Chem 88:2697–2702
8. Guldi DM, Mody TD, Gerasimchuk NN, Magda D, Sessler JL (2000) Influence of large metal cations on the photophysical properties of texaphyrin, a rigid aromatic chromophore. J Am Chem Soc 122:8289–8298
9. Nakanishi T, Suzuki Y, Doi Y, Seki T, Koizumi H, Fushimi K, Fujita K, Hinatsu Y, Ito H, Tanaka K, Hasegawa Y (2014) Enhancement of optical faraday effect of nonanuclear Tb (III) complexes. Inorg Chem 53:7635–7641
10. Casanova D, Llunell M, Alemany P, Alvarez S (2005) The rich stereochemistry of eight-vertex polyhedra: a continuous shape measures study. Chem Eur J 11:1479–1494
11. Pinsky M, Avnir D (1998) Continuous symmetry measures. 5. The classical polyhedra. Inorg Chem 37:5575–5582
12. SHAPE, version 2.1 (2010) Continuous shape measures calculations. Electronic Structure Group, Universitat de Barcelona, Spain
13. Xu J, Radkov E, Ziegler M, Raymond KN (2000) Plutonium (IV) Sequestration: Structural and Thermodynamic Evaluation of the Extraordinarily Stable Cerium (IV) Hydroxypyridinonate Complexes1. Inorg Chem 39:4156–4164
14. Omagari S, Nakanishi T, Kitagawa Y, Seki T, Fushimi K, Ito H, Meijerink A, Hasegawa Y (2018) Spin-orbit coupling dependent energy transfer in luminescent nonanuclear Yb-Gd/Yb-Lu clusters. J Lumin 201:170–175

15. Hasegawa Y, Ohkubo T, Nakanishi T, Kobayashi A, Kato M, Seki T, Ito H, Fushimi K (2013) Effect of ligand polarization on asymmetric structural formation for strongly luminescent lanthanide complexes. Eur J Inorg Chem 5911–5918
16. Miyata K, Nakagawa T, Kawakami R, Kita Y, Sugimoto K, Nakashima T, Harada T, Kawai T, Hasegawa Y (2011) Remarkable luminescence properties of lanthanide complexes with asymmetric dodecahedron structures. Chem Eur J 17:521–528
17. El-Sayed MA (1963) Spin—Orbit coupling and the radiationless processes in nitrogen heterocyclics. J Chem Phys 38:2834–2838
18. Bünzli J-CG, Piguet C (2005) Taking advantage of luminescent lanthanide ions. Chem Soc Rev 34:1048–1077

# Chapter 5
# Back Energy Transfer in Nonanuclear Terbium Clusters

**Abstract** Energy transfer from organic ligands to a lanthanide ion is the main concept of sensitized luminescence in lanthanide complexes. Back energy transfer is the reverse process, which occur when the energy of the triplet excited state is close to the emitting excited state of the lanthanide ion and can lead to decrease in sensitization efficiency and therefore the quantum yield as well. Generally, the back energy transfer rate constant is proportional to the energy gap between the triplet excited state and the emitting excited state of the lanthanide, but there are some compounds that deviate from this trend. This chapter focuses on $[Tb_9(\mu\text{-}OH)_{10}(L)_{16}]NO_3$ where ligand L is a derivative of methyl salicylate ligands to change the triplet excited state energy without significantly changing the structure so that direct comparison of back energy transfer rate constant can be performed. The conclusion of this chapter is that back energy transfer rate constant is determined not only by the energy gap but also by the activation energy and the frequency factor (electronic correlation) of the process.

## 5.1 Introduction

Sensitization efficiency in Ln(III) complexes is one of the most critical factors that affect the quantum yield of Ln(III) complexes [1, 2]. Chapter 4 described the effect of SOC on the photophysical processes within the ligands of Ln(III) complexes that affects the sensitization efficiency. In this chapter, the FET process is the primary focus. FET efficiency is known to be dependent on the $T_1$ state energy in relative to the excited 4f-state of a Ln(III) ion [1, 3]. In particular, $T_1$ energy of Tb(III) complexes has a strong dependence on the quantum yield (and emission lifetimes) due to BET as mentioned in Chap. 1 (Sect. 1.3.1) [4]. Within the region of the $T_1$ state energy below 1850 $cm^{-1}$, some of the Tb(III) complexes deviate from such trend (emission lifetimes as well). This is an important fact because it implies that there is a factor(s) other than the $T_1$-$Tb(^5D_4)$ energy gap that affects the BET.

Herein, nonanuclear Tb(III) clusters were used for the investigation of the mechanism of BET [5]. The main objective is to derive the parameters that affect the BET rate constant. By employing methyl salicylate with methyl substituent on the aro-



S. Omagari, *Energy Transfer Processes in Polynuclear Lanthanide Complexes*, Springer Theses, https://doi.org/10.1007/978-981-13-6049-7_5

matic part as a ligand, the excited state energies of the ligand can be altered. Methyl is a weak electron-donating substituent with only of inductive effect (no resonance effect), allowing only a slight modification to the electronic structure and the excited state energies of the ligand. The reason for using nonanuclear Tb(III) clusters is similar to that explained in Sect. 4.1. The structure of the cluster core is likely maintained. Since the luminescence property of Ln(III) ions is sensitive to the crystal field (see Sect. 1.2.2) [6–9], maintaining the core structure is necessary for direct comparison of the effects caused by the different ligands. However, there is a difference in the modifications mentioned in Chap. 4 and those in this chapter. In Chap. 4, different Ln(III) ions were mixed within the cluster whereas, in this chapter, different ligands are used. Therefore, the effect of the ligands on the luminescent properties of the cluster core must first be elucidated.

This chapter is composed of two parts: (1) elucidation of the effect of intramolecular interactions of the ligands on the photophysical properties of the clusters, and (2) investigation of the mechanism of BET and factors that determine the BET rate constants. The organic ligands used for this study are provided in Fig. 5.1. Alkylsalicylate ligands of different alkyl chain length, namely ethyl (Et), propyl (Pr), butyl (Bu), and hexyl (He), which changes the intramolecular interaction between ligands, were employed for (1). Methylsalicylate ligands (Me) with methyl substituent on the 4-position (4Me) and 5-position (5Me) of the aromatic part were used for (2). The synthesized nonanuclear Tb(III) clusters with these ligands above ($[Tb_9(\mu\text{-}OH)_{10}(L)_{16}]NO_3$, where L = Me, 4Me, 5Me, Et, Pr, Bu, or He) were identified by IR, mass spectroscopy, and elemental analysis. The photophysical properties were investigated by emission and absorption spectroscopy, absolute quantum yield measurements, and emission lifetimes.

## 5.2 Experimental Section

### *5.2.1 Material*

Salicylic acid ($C_6H_4(OH)COOH$), methanol ($CH_3OH$), ethanol ($C_2H_5OH$), propanol ($C_3H_7OH$), n-butanol ($C_4H_9OH$), Tb(III) nitrate hexahydrate ($Tb(NO_3)_3 \cdot 6H_2O$) and triethylamine ($C_6H_{15}N$) were purchased from Kanto Chemical Co. Inc. Gd(III) nitrate hexahydrate ($Gd(NO_3)_3 \cdot 6H_2O$) were purchased from Sigma-Aldrich Co. Inc. n-hexanol ($C_6H_{13}OH$), 4-methyl-methylsalicylate ($C_{11}H_{14}O_3$) and 5-methyl-methylsalicylate ($C_{11}H_{14}O_3$) were purchased from Tokyo Chemical Industry Co., LTD. All other chemicals and solvents were reagent grade and were used without further purification.

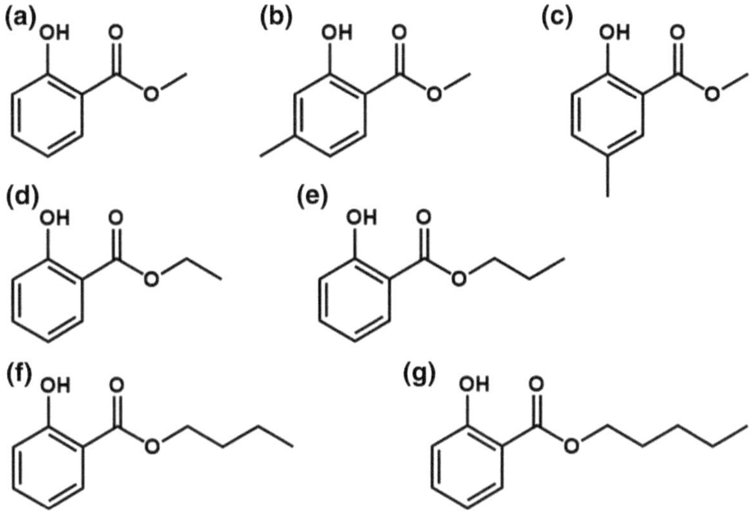


**Fig. 5.1** Organic ligands used for the study described in this chapter. **a** Me, **b** 4Me, **c** 5Me, **d** Et, **e** Pr, **f** Bu, **g** He

### *5.2.2 Apparatus*

Infrared spectra were recorded on a ThermoNicolet Avatar 320 FTIR spectrometer. XRD spectra were characterized by a RIGAKU X-ray diffractometer SmartLab. FAB-MS spectra were measured on a JEOL JMS-700TZ. Elemental analyses were performed by a MICRO CORDER JM10.

### *5.2.3 Synthesis*

**General synthetic procedures of $Tb_9$ clusters with long alkyl chain** The general synthetic scheme is provided in Chap. 4 and will be omitted in this section. Appropriate alkyl salicylate ligand (2.7 mmol) was dissolved in methanol, and triethylamine (0.44 mL, 4.40 mmol) was added to this solution while stirred at 40 °C. $Tb(NO_3)_3 \cdot 6H_2O$ (0.686 g, 1.52 mmol) dissolved in methanol was added dropwise to this solution and further stirred for 40 min. The resulting white precipitate (the target cluster) was vacuum filtered and dried in vacuum. The clusters were recrystallized by evaporation method from a concentrated methanol solution.

**General synthetic procedures of $Ln_9$ clusters with methyl-methylsalicylate ligands** The synthetic procedures are the same as the procedures above but with a different scale. Appropriate methylsalicylate ligand (5.40 mmol) was dissolved in methanol, and triethylamine (1.22 ml, 8.80 mmol) was added to this solution while stirred at 40 °C. $Ln(NO_3)_3 \cdot 6H_2O$ (3.04 mmol) dissolved in methanol was added dropwise to this solution and further stirred for 40 min. The resulting white precipitate (the target cluster) was vacuum filtered and dried in vacuum.

**$Tb_9$-Et ($[Tb_9(\mu\text{-}OH)_{10}(Et)_{16}]^+NO_3^-$)**: Elemental analysis calculated for $C_{144}H_{154}NO_{61}Tb_9$: C, 40.17%, H, 3.61%, N, 0.33%. Found: C, 39.91%, H, 3.59%, N, 0.34%. FAB-MS: m/z = 4242.2 ($[Tb_9(\mu\text{-}OH)_{10}(Et)_{16}]^+$).
**$Tb_9$-Pr ($[Tb_9(\mu\text{-}OH)_{10}(Pr)_{16}]^+NO_3^-$)**: Elemental analysis calculated for $C_{160}H_{186}NO_{61}Tb_9$: C, 42.43%, H, 4.14%, N, 0.31%. Found: C, 42.12%, H, 4.10%, N, 0.34%. FAB-MS: m/z = 4466.6 ($[Tb_9(\mu\text{-}OH)_{10}(Pr)_{16}]^+$).
**$Tb_9$-Bu ($[Tb_9(\mu\text{-}OH)_{10}(Bu)_{16}]^+NO_3^-$)**: Elemental analysis calculated for $C_{176}H_{218}NO_{61}Tb_9$: C, 44.47%, H, 4.62%, N, 0.29%. Found: C, 44.16%, H, 4.57%, N, 0.51%. FAB-MS: m/z = 4691.3 ($[Tb_9(\mu\text{-}OH)_{10}(Bu)_{16}]^+$).
**$Tb_9$-He ($[Tb_9(\mu\text{-}OH)_{10}(He)_{16}]^+NO_3^-$)**: Elemental analysis calculated for $C_{208}H_{282}NO_{61}Tb_9$: C, 48.02%, H, 5.46%, N, 0.27%. Found: C, 47.8%, H, 5.28%, N, 0.31%. FAB-MS: m/z = 5140.0 ($[Tb_9(\mu\text{-}OH)_{10}(He)_{16}]^+$).
**$Tb_9$-Me ($[Tb_9(\mu\text{-}OH)_{10}(Me)_{16}]^+NO_3^-$)**: Selected IR (ATR, $cm^{-1}$): 3573 (w, O–H), 3235 (w, O–H), 2956 (m, C–H), 1319 (s, C–O). FAB-MS: m/z = 4018.8 ($[Tb_9(\mu\text{-}OH)_{10}(Me)_{16}]^+$).
**$Tb_9$-4Me ($[Tb_9(\mu\text{-}OH)_{10}(4Me)_{16}]^+NO_3^-$)**: Selected IR (ATR, $cm^{-1}$): 3573 (w, O–H), 3235 (w, O–H), 2956 (m, C–H), 1319 (s, C–O). FAB-MS: m/z = 4242.1 ($[Tb_9(\mu\text{-}OH)_{10}(4Me)_{16}]^+$). Elemental analysis calculated for $C_{148}H_{160}NO_{61}Tb_9$: C, 40.17%, H, 3.61%, N, 0.33%. Found: C, 39.45%, H, 3.66%, N, 0.31%.
**$Tb_9$-5Me ($[Tb_9(\mu\text{-}OH)_{10}(5Me)_{16}]^+NO_3^-$)**: Selected IR (ATR, $cm^{-1}$): 3573 (w, O–H), 3235 (w, O–H), 2956 (m, C–H), 1319 (s, C–O). FAB-MS: m/z = 4242.6 ($[Tb_9(\mu\text{-}OH)_{10}(5Me)_{16}]^+$). Elemental analysis calculated for $C_{148}H_{160}NO_{61}Tb_9$: C, 40.17%, H, 3.61%, N, 0.33%. Found: C, 39.40%, H, 3.69%, N, 0.35%.
**$Gd_9$-Me ($[Gd_9(\mu\text{-}OH)_{10}(Me)_{16}]^+NO_3^-$)**: Selected IR (ATR, $cm^{-1}$): 3573 (w, O–H), 3235 (w, O–H), 2956 (m, C–H), 1319 (s, C–O). FAB-MS: m/z = 4002.0 ($[Gd_9(\mu\text{-}OH)_{10}(Me)_{16}]^+$).
**$Gd_9$-4Me ($[Gd_9(\mu\text{-}OH)_{10}(4Me)_{16}]^+NO_3^-$)**: Selected IR (ATR, $cm^{-1}$): 3573 (w, O–H), 3235 (w, O–H), 2956 (m, C–H), 1319 (s, C–O). FAB-MS: m/z = 4228.2 ($[Gd_9(\mu\text{-}OH)_{10}(4Me)_{16}]^+$).
**$Gd_9$-5Me ($[Gd_9(\mu\text{-}OH)_{10}(5Me)_{16}]^+NO_3^-$)**: Selected IR (ATR, $cm^{-1}$): 3573 (w, O–H), 3235 (w, O–H), 2956 (m, C–H), 1319 (s, C–O). FAB-MS: m/z = 4229.6 ($[Gd_9(\mu\text{-}OH)_{10}(5Me)_{16}]^+$).

### *5.2.4 Spectroscopy*

UV/Vis absorption spectra ($1.0 \times 10^{-4}$ M chloroform solution) and diffuse reflectance (powdered samples) were recorded using a JASCO V-670 spectrometer. Emission spectra were recorded using a Horiba FluoroLog-3 ps. Emission lifetimes were measured using the third harmonics (355 nm) of a Q-switched Nd:YAG laser (Spectra Physics, INDI-50, FWHM = 5 ns, λ = 1064 nm) and a photomultiplier (Hamamatsu photonics, R928). The Nd:YAG laser response was monitored with a digital oscilloscope (Sony Tektronix, TDS3052, 500 MHz) synchronized to the single-pulse excitation. Emission lifetimes were determined from the slope of logarithmic plots of the decay profiles.

The absolute emission quantum yields were estimated using a JASCO FP-6600 spectrometer attached with a JASCO ILF-533 integrating sphere unit ($\varphi = 100$ mm). The wavelength dependences of the detector response and the beam intensity of Xe light source for each spectrum were calibrated using a standard light source. Quantum yield can be calculated by [$\Phi = (I_c - I_a)/(B_a - B_c)$] with $I_c$ being the integrated emission spectrum of the sample, $I_a$ the integrated "blank" emission spectrum, $B_a$ the "blank" absorption, and $B_c$ the sample absorption at the excitation wavelength.

### 5.2.5 *Computational Details*

DFT geometry optimizations and MO calculations of simplified clusters were performed with Gaussian 09 (Revision B.01) by employing the three-parameter hybrid functional of Becke based on the correlation functional of Lee, Yang, and Parr (B3LYP). The 6-31G(d) basis set was used for all atoms.

## 5.3 Results and Discussion

### 5.3.1 *Structure*

Identification of the $Tb_9$ clusters was performed by fast-atomic bombardment mass spectroscopy (FAB-MS). The m/z for $Tb_9$-Me, $Tb_9$-4Me, $Tb_9$-5Me, $Tb_9$-Et, $Tb_9$-Pr, $Tb_9$-Bu, and $Tb_9$-He were determined to be 4018.8, 4242.1, 4242.6 4242.2, 4466.6, 4691.3, and 5140.0, respectively. The obtained m/z can be attributed to $Tb_9$ clusters without the $NO_3^-$ counter-anion. The basic molecular structures of $Ln_9$ clusters are provided in Chap. 4 (see Sect. 4.3.1). The single crystal X-ray analyses of the $Tb_9$-Me, $Tb_9$-Bu, and $Tb_9$-He, in particular, are already reported [5, 10]. Given the mass spectrometry results, the other clusters without single crystal X-ray analysis also have the same basic structure. In Chap. 4, it was mentioned that the $Ln_9$ clusters possess nine oxygen-bridged Ln(III) ions with an "hour-glass" structure surrounded by sixteen alkyl salicylate ligands. Focusing on the packing structure of the $Tb_9$ clusters ($Tb_9$-He as an example taken from Ref. [7]), a chain of intramolecular CH–π and π–π interactions between ligands can be seen (Fig. 5.2). The π–π interactions are observed between two aromatic rings (center-to-center distance of ~4.25 Å) around the equator of the cluster. The CH–π interactions are also found between alkyl chain and aromatic ring (>2.75 Å) near the top and bottom of the cluster. From

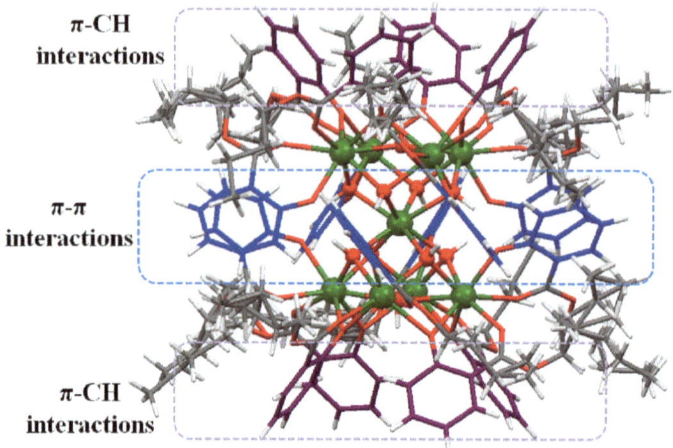


**Fig. 5.2** Intramolecular CH–π and π–π interactions between ligands in $Tb_9$-He. Reprinted from Ref. [15] with permission of the Japan Society of Vacuum and Surface Science

the structure, the methyl substituent on the aromatic part of methylsalicylate ligands (4Me and 5Me) would face outside of the cluster with very small steric hindrance. This is likely the reason behind the retained structure of $Tb_9$-4Me and $Tb_9$-5Me as nonanuclear Ln(III) clusters.

### *5.3.2 Intramolecular Interactions and Photophysical Properties*

In order to investigate the influence of intramolecular interaction on the photophysical properties of the organic ligands, absorption spectra of Et, Pr, Bu, He, and the $Tb_9$ clusters with these ligands in $1.0 \times 10^{-5}$ M chloroform solution were measured. The spectra of the ligands and the clusters are shown in Figs. 5.3a and b, respectively. All ligands showed a negligibly small difference in absorption spectra. Meanwhile, a difference in absorption coefficient $\varepsilon$ is observed for the $Tb_9$ clusters with the different alkyl chain, although there is no proportional relationship between the length of alkyl chain and the absorption coefficient. $Tb_9$-Bu showed the largest absorption coefficient of 55508 $M^{-1}$ $cm^{-1}$ while $Tb_9$-Et showed the smallest absorption coefficient of 46432 $M^{-1}$ $cm^{-1}$. The change in the oscillator strength is suggested to be caused by changes in the intramolecular interactions. However, given the lack of proportional relationship between the length of alkyl chain and the absorption coefficient, the effect is complicated and is not easily predictable.

The influence of intramolecular interactions on the photophysical properties of the Tb(III) ions in the clusters was investigated by the emission spectra of the $Tb_9$ clusters. The emission of Ln(III) ions is known to be sensitive to the surrounding

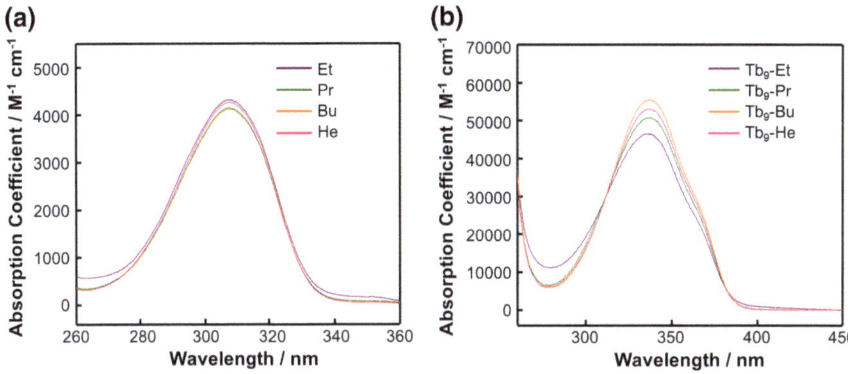


**Fig. 5.3** Absorption spectra of **a** the ligands dissolved in $1.0 \times 10^{-4}$ M chloroform solution, and **b** the clusters in $1.0 \times 10^{-5}$ M chloroform solution

environment in terms of spectral shape (Stark-splitting) and radiative rate constant [6–8]. The emission spectra of $Tb_9$-Et, $Tb_9$-Pr, $Tb_9$-Bu, and $Tb_9$-He in $1.0 \times 10^{-4}$ M chloroform solution are shown in Fig. 5.4a. The clusters were excited at 375 nm. The emission spectra were normalized at the $^5D_4 \rightarrow {}^7F_5$ emission. The emission bands of the $Tb_9$ clusters were observed at approximately 487, 548, 582, 623, 651, and 681 nm, which are attributed to the typical 4f-4f transition of Tb(III) ions ($^5D_4 \rightarrow {}^7F_J$, where $J = 6 - 1$). The spectral shape was identical for all the clusters, indicating that the clusters have identical coordination environment despite the difference in the intramolecular interactions. The identical spectral shapes also imply that the radiative rate constants are the same for these clusters.

The emission decay profiles and the emission lifetimes $\tau_{obs}$ of $Tb_9$-Et, $Tb_9$-Pr, $Tb_9$-Bu, and $Tb_9$-He in $1.0 \times 10^{-4}$ M chloroform solution are shown in Fig. 5.4b and Table 5.1, respectively. $Tb_9$-Et had the longest lifetime of 1.26 ms. $Tb_9$-Pr and $Tb_9$-Bu showed similar lifetime of 1.13 and 1.17 ms, respectively. $Tb_9$-He exhibited the shortest lifetime of 1.00 ms. Since the radiative rate constants of these clusters are likely the same, the difference in the lifetimes likely arose from the difference in the nonradiative rate constants. The quantum yields of the clusters by ligand excitation at 375 nm ($\Phi_{\pi\pi*}$) are around 40% (Table 5.1). The emission lifetime and the quantum yield results altogether imply that the increase in the length of the alkyl chain leads to the increase in the nonradiative rate constant of Tb(III) ions, but also in the sensitization efficiency ($\eta_{sens}$) as well.

It should be note here that the quantum yields of the clusters by Tb(III) ion excitation at 487 nm ($\Phi_{ff}$) were also measured, which in principle allow the calculation of the radiative and nonradiative rate constants ($kr$ and $knr$, respectively). The values are summarized in Table 5.1. However, the small absorption coefficient of the Tb(III) ions (approximately 10 $M^{-1}$ $cm^{-1}$ for a cluster containing nine Tb(III) ions) and thus the very low emission intensity observed in the absolute quantum yield measurement makes these values unreliable (see Fig. 5.9 in Sect. 5.5.1). Moreover, there exist two different coordination sites in the clusters in these clusters as mentioned in 4.3.1.

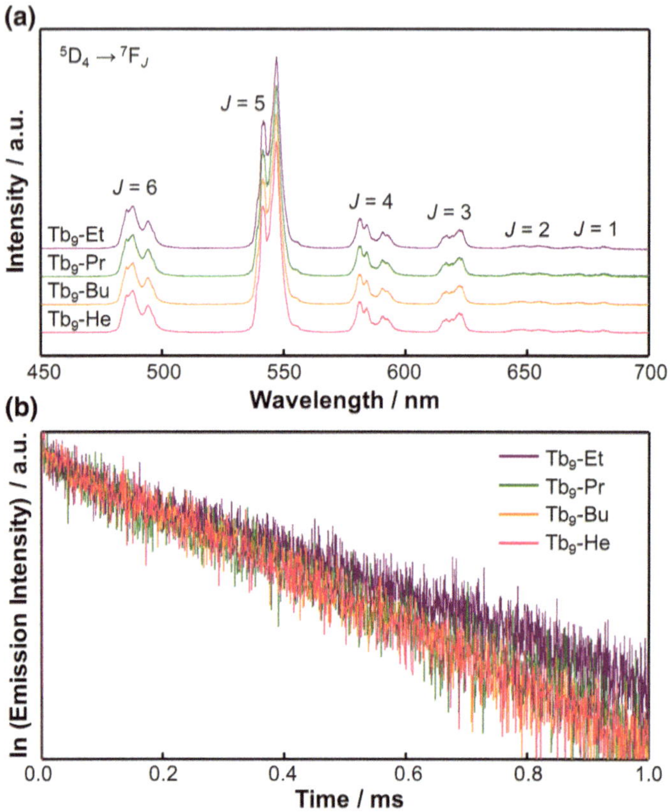


**Fig. 5.4** **a** Emission spectra of Tb9-Et, Tb9-Pr, Tb9-Bu, and Tb9-He in $1.0 \times 10^{-4}$ M chloroform solution ($\lambda_{EX} = 375$ nm). The emission spectra were normalized and given an offset. **b** Emission decay profiles of the clusters in $1.0 \times 10^{-4}$ M chloroform solution

**Table 5.1** Photophysical properties of Tb9-Et, Tb9-Pr, Tb9-Bu, and Tb9-He

| Cluster | $\varepsilon$/M$^{-1}$ cm$^{-1}$ | $\Phi_{\pi\pi^*}$ | $\Phi_{ff}^{a}$ | $\eta_{sens}^{a}$ | $\tau_{obs}$/ms | $\tau_{rad}$/ms$^{a}$ | $k_r$/s$^{-1a}$ | $k_{nr}$/s$^{-1a}$ |
|---|---|---|---|---|---|---|---|---|
| Tb9-Et | 46432 | 0.42 | 0.69 | 0.61 | 1.26 | 1.82 | 548 | 246 |
| Tb9-Pr | 50665 | 0.40 | 0.56 | 0.72 | 1.13 | 2.03 | 493 | 392 |
| Tb9-Bu | 55508 | 0.39 | 0.42 | 0.93 | 1.17 | 2.79 | 358 | 497 |
| Tb9-He | 52897 | 0.42 | 0.86 | 0.49 | 1.00 | 1.16 | 859 | 137 |

Redrawn from Ref. [15] with permission of the Japan Society of Vacuum and Surface Science
[a]These values are unreliable due to the uncertainties in the measurement of $\Phi_{ff}$

When ligands are excited first, the Tb(III) ions coordinated directly to the ligands will be excited first and then the center Tb(III) ion that is not coordinated to alkyl salicylate ligands. On the other hand, when Tb(III) ions are directly excited, all of the Tb(III) ions are excited at first. This fundamental difference in the initial excitation of Tb(III) ions renders the $\Phi_{\pi\pi*}$ and $\Phi_{ff}$ to be a (directly) incompatible values. This will be explained in Chap. 6 in details.

Provided with the results above, it is suggested that the intramolecular interactions mainly affect the photophysical properties on the ligand side than the cluster core side. The effect of ionic radii of Ln(III) ions on the photophysical property of the clusters was also briefly mentioned in Chap. 4 where the Yb(III) emission was monitored when the cluster was mixed with Gd(III) and Lu(III) ions. It was found that ionic radii do not affect the radiative rate constant of Yb(III) ions, but only the nonradiative rate constant. Overall, the nonanuclear Ln(III) clusters contain a rigid core and maintain the fundamental photophysical properties of Ln(III) ions. Such property of the $Tb_9$ clusters sets the stage for the next step where the influence of using methyl salicylate derivatives on the energy transfer rate constant is investigated. It is preferred for such investigation that the fundamental properties of Tb(III) ions core are maintained as much as possible so that the effect arising from the ligands can be directly compared.

### *5.3.3 Mechanism of Back Energy Transfer*

The effect of the methyl substituent on the photophysical properties of methyl salicylate ligands was investigated by quantum chemical calculations (DFT with B3LYP functional). The calculations were performed for Me, 4Me, and 5Me ligands bonded with Li(I) ion to simplify the discussion. These complexes will be referred to as Li-Me, Li-4Me, and Li-5Me. The highest-occupied and lowest-unoccupied molecular orbitals (HOMO and LUMO, respectively) of Li-Me, Li-4Me, and Li-5Me are depicted in Fig. 5.5a, and their energies are summarized in Fig. 5.5b. Focusing on Li-Me, both the HOMO and LUMO are distributed throughout the aromatic part of Me ligand including the carboxyl part of the ester group. The HOMO was poorly distributed on the 4-position carbon of the aromatic ring (opposite side of the ester group) while LUMO was poorly distributed on the 5-position. Li-4Me and Li-5Me did not show much difference in the distribution of the HOMO and LUMO compared to those of Li-Me, but the energies of these two orbitals were different. In Li-4Me, the LUMO energy was higher than that of Li-Me (0.095 eV) while the HOMO did not show much shift (0.044 eV). The opposite was observed for Li-5Me where the HOMO rose in energy (0.136 eV) while the LUMO only slightly shifted (0.035 eV). Li-4Me and Li-5Me showed the opposite effect regarding which orbital energy rises because of the distribution differences of HOMO and LUMO on 4- and 5-positions. The resulting HOMO-LUMO energy gaps of Li-Me, Li-4Me, and Li-5Me were 4.14, 4.19, and 4.04 eV, respectively.

The rise in the energy is caused by the methyl substituent, which donates the electron into HOMO or LUMO leading to the rise in the electron density and thus the

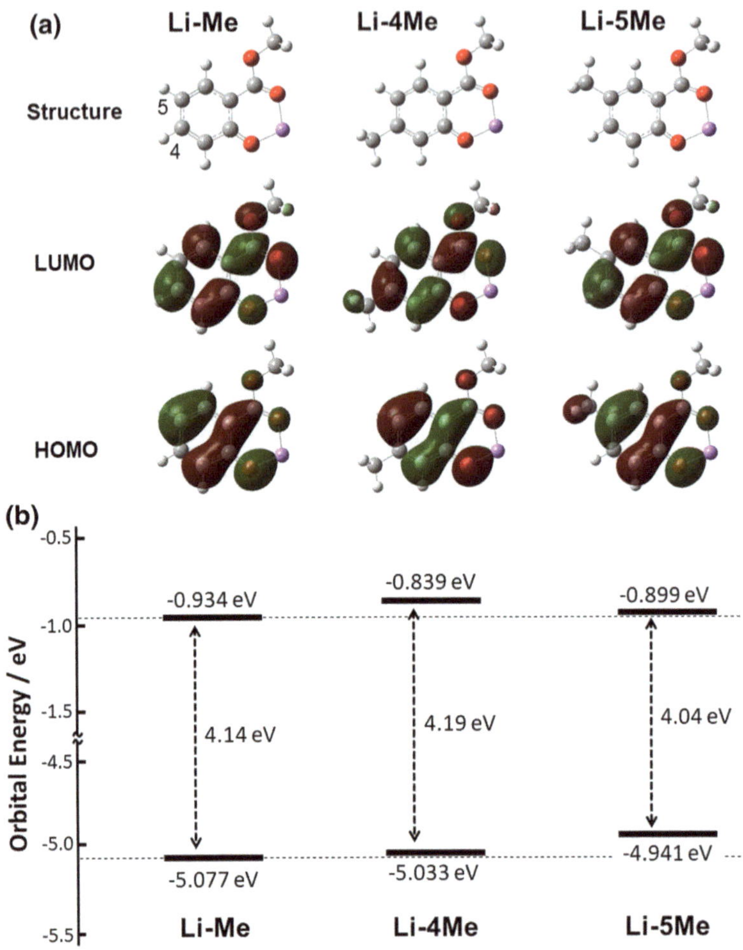


**Fig. 5.5** **a** Molecular orbitals of Li-Me, Li-4Me, and Li-5Me calculated by DFT (B3LYP functional with 6-31G(d) basis set). **b** Energy diagram of the orbital energy of HOMO and LUMO. Dotted horizontal lines are the HOMO and LUMO energy level of Li–Me

energy as well. Methyl substituent is a weak electron donor with only the inductive effect. This shows that using methyl substituent is an effective method for slightly modifying the HOMO-LUMO gap without significantly altering the electronic structure. The clusters are composed of 16 methylsalicylate ligands, and the result above is not necessarily accurate in predicting the effect of the methyl substituent because of the intramolecular interaction between the ligands as explained in 5.3.2. Nonetheless, the goal here is to slightly modify the energy gap between the $T_1$ and $S_0$ states so that the FET process is changed. The calculation results justify the use of these ligands to investigate the effect of electronic energy on the energy transfer processes.

The diffuse reflectance spectra of $Tb_9$-Me, $Tb_9$-4Me, and $Tb_9$-5Me were carried out to compare the absorption edge of the ligands (Fig. 5.6a). The clusters were measured as a powdered state because of the poor solubility. The absorption edge slightly shifted toward the longer wavelength in the order of $Tb_9$-Me, $Tb_9$-4Me, and $Tb_9$-5Me. TD-DFT calculation of Li-Me, Li-4Me, and Li-5Me showed that the $S_0$-$S_1$ transition of these molecules are of HOMO-to-LUMO origin ($\pi$–$\pi^*$ transition). The absorption of the clusters can be assigned to the $\pi$–$\pi^*$ transition of the methylsalicylate ligands.

Figure 5.6b shows the emission spectra of $Gd_9$-Me, $Gd_9$-4Me, and $Gd_9$-5Me excited at 380 nm at 100 K. Gd(III) ions have excited states above 32000 $cm^{-1}$, which is well above of the ligands and energy transfer do not proceed from the

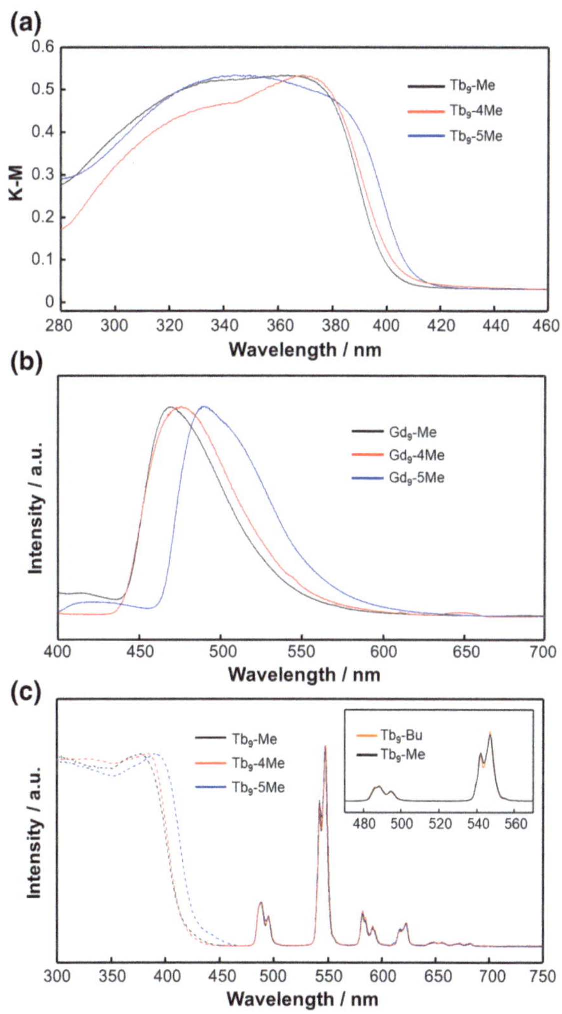


**Fig. 5.6** **a** Diffuse reflectance spectra of $Tb_9$–Me, $Tb_9$–4Me, and $Tb_9$–5Me. **b** Emission spectra of $Gd_9$–Me, $Gd_9$–4Me, and $Gd_9$-5Me at 100 K. **c** Emission spectra of $Tb_9$-Me, $Tb_9$-4Me, and $Tb_9$-5Me at room temperature. The inset shows the comparison of spectral shape between $Tb_9$-Bu and $Tb_9$-Me. The measurement was performed as powder, and the excitation wavelength was 380 nm for the emission spectra

ligands to Gd(III) ions. As a result, emission from the ligands is observed, and at lower temperature phosphorescence from the triplet excited state. This is a common method to identify the $T_1$ state energy of Ln(III) complexes [3, 4]. These result clearly indicate a gradual shift to longer wavelength in the order of $Gd_9$-Me, $Gd_9$-4Me, and $Gd_9$-5Me, similar to the diffusion reflectance spectra results. The energy of the $T_1$ state ($T_1$-$S_0$ gap) of $Gd_9$-Me, $Gd_9$-4Me, and $Gd_9$-5Me were estimated to be 21277, 21053, and 20408 $cm^{-1}$, respectively. The combination of diffuse reflectance spectra of the $Tb_9$ clusters and emission spectra of the $Gd_9$ clusters confirms that the slight modification of the (singlet and triplet) excited states of the ligands using methyl substituents was successful.

The steady-state emission spectra of $Tb_9$-Me, $Tb_9$-4Me, and $Tb_9$-5Me in solid-state excited at 380 nm are shown in Fig. 5.6c. The clusters were poorly soluble in solvents and had to be performed in the solid state. The spectral shape of Tb(III) emission of these clusters was identical with those of $Tb_9$-Et, $Tb_9$-Pr, $Tb_9$-Bu, and $Tb_9$-He (Fig. 5.6c inset). Therefore, coordination geometry of Tb(III) ions is once again similar among all of the clusters mentioned in this chapter. The excitation spectra of the clusters monitored at 548 nm also showed a shift to the longer wavelength region in the order of $Tb_9$-Me, $Tb_9$-4Me, and $Tb_9$-5Me, which are consistent with the diffuse reflection spectra (Fig. 5.6a) and triplet excited state phosphorescence spectra (Fig. 5.6b). The result further supports that Ln(III) ions core of the cluster is rigid and retains its core structure even with changes in the intramolecular interactions. From the emission spectra of $Tb_9$-Me, $Tb_9$-4Me, and $Tb_9$-5Me as well as their Gd(III) ion analogs, the energy gaps between the $T_1$ state and the $^5D_4$ state of the Tb(III) ions in the clusters $\Delta E$ $(= E_{T_1} - E_{^5D_4})$ are summarized in Table 5.2. $Tb_9$-Me showed the largest $\Delta E$ of 740 $cm^{-1}$ while $Tb_9$-5Me showed the smallest $\Delta E$ of-130 $cm^{-1}$. Generally, the sensitization efficiency is known to be proportional to $\Delta E$ especially when the value is less than 1850 $cm^{-1}$ since BET is known to occur under this energy gap [4, 11]. The clusters therefore likely show the effect of BET.

The emission decay profiles (excited at 355 nm) and the absolute quantum yields (excited at 380 nm) of the clusters were measured. The values are summarized in Table 5.2. The profiles and its value $\tau_{obs}$ of the clusters revealed single-exponential decays as shown in Fig. 5.7a. The largest $\tau_{obs}$ was observed for $Tb_9$-4Me (815 μs), while the smallest was observed for $Tb_9$-5Me (190 μs). $Tb_9$-Me showed a value in the middle (264 μs). The emission quantum yields showed a similar trend where the values were in the order from large to small $Tb_9$-4Me ($\Phi_{\pi\pi*} = 31\%$), $Tb_9$-Me

**Table 5.2** Photophysical properties of $Tb_9$-Me, $Tb_9$-4Me, and $Tb_9$-5Me

| Cluster | $\Delta E/cm^{-1}$ | $\tau_{obs}/\mu s$ | $Ea/cm^{-1}$ | $D/s^{-1}$ | $\Phi_{\pi\pi*}$ |
|---|---|---|---|---|---|
| $Tb_9$-Me | +740 | 284 | 1822 | $5.76 \times 10^5$ | 0.07 |
| $Tb_9$-4Me | +520 | 815 | 2082 | $1.17 \times 10^6$ | 0.31 |
| $Tb_9$-5Me | −130 | 190 | 1496 | $1.17 \times 10^6$ | 0.02 |

Reprinted from Ref. [16] with permission from the American Chemical Society

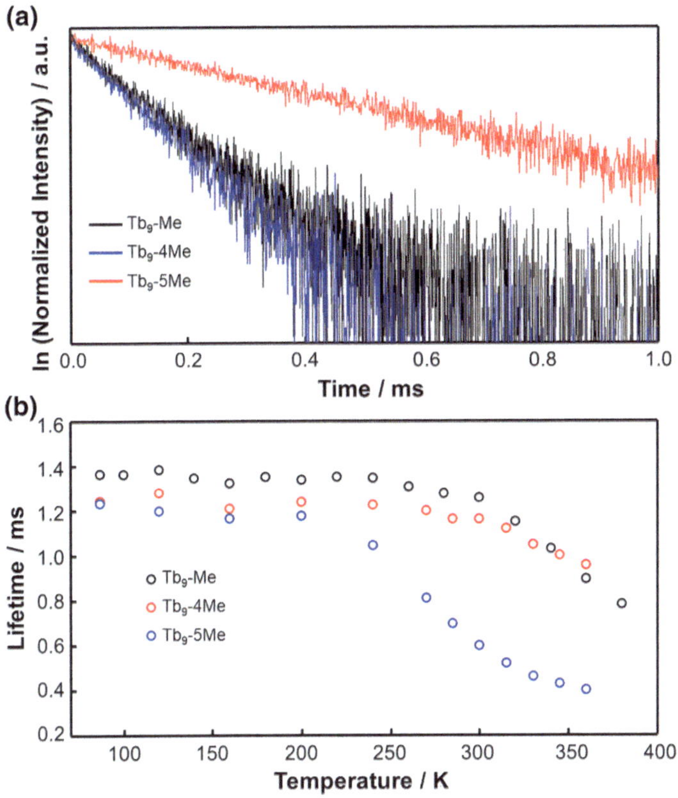


**Fig. 5.7** **a** Emission decay profiles of Tb9-Me, Tb9-4Me, and Tb9-5Me at room temperature. **b** Temperature dependence of emission lifetimes of the clusters. All measurements were performed as powder and excited at 355 nm

($\Phi_{\pi\pi*} = 6.7\%$), and Tb9-5Me ($\Phi_{\pi\pi*} = 2.4\%$). These results are inconsistent with the trend observed for the energy gap $\Delta E$.

In order to explain this inconsistency, the focus is now primarily on the BET rate constant $k_{\mathrm{BET}}$, which would directly provide information on the energy transfer processes. BET can be considered as a temperature dependent process because the process needs to be assisted by thermal energy to compensate for the energy difference between $^5D_4$ state and $T_1$ state. Ignoring temperature dependence of non-radiative rate constant of Tb(III) ions, at sufficiently low temperature, BET does not proceed. The BET rate constant $k_{\mathrm{BET}}$ can be expressed using emission lifetimes by:

$$k_{\mathrm{BET}} = (kr + knr + k_{\mathrm{BET}}) - (kr + knr) = \frac{1}{\tau(T)} - \frac{1}{\tau(87\mathrm{K})}, \quad (5.1)$$

where $kr$ and $knr$ are radiative and nonradiative rate constants, respectively. $\tau(T)$ is the emission lifetime at a given temperature $T$, and $\tau(87\mathrm{K})$ is the emission lifetime at 87 K. Since this process is temperature dependent, Arrhenius plot can be applied [12, 13] (see Sects. 5.5.2 and 5.5.3 for details):

$$k_{\mathrm{BET}} = D\exp\left(-\frac{Ea}{k_{\mathrm{B}}T}\right), \tag{5.2}$$

where D is the frequency factor. By combining Eqs. (5.1) and (5.4), the following equation is derived:

$$k_{\mathrm{BET}} = \frac{1}{\tau(T)} - \frac{1}{\tau(87\mathrm{K})} = \mathrm{D}\exp\left(-\frac{Ea}{k_{\mathrm{B}}T}\right). \tag{5.3}$$

The equation implies that by plotting $\frac{1}{\tau(T)} - \frac{1}{\tau(87\mathrm{K})}$ against temperature and fitting the right side of the equation leads to the derivation of the frequency factor $D$ and the activation energy of BET $Ea$.

The temperature dependence (87–380 K) of emission lifetimes for $Tb_9$-Me, $Tb_9$-4Me, and $Tb_9$-5Me are provided in Fig. 5.7b. The derived frequency factor $D$ and the activation energy $Ea$ are summarized in Table 5.2. The activation energy $Ea$ for $Tb_9$-Me, $Tb_9$-4Me, and $Tb_9$-5Me were 1822 $cm^{-1}$, 2082 $cm^{-1}$, and 1496 $cm^{-1}$, respectively. The values were clearly larger than the energy gap $\Delta E$ by more than 1000 $cm^{-1}$. Furthermore, the frequency factor $D$ for $Tb_9$-Me, $Tb_9$-4Me, and $Tb_9$-5Me were $5.76 \times 10^5$, $1.17 \times 10^6$, and $1.17 \times 10^6\ s^{-1}$, respectively. These value of the frequency factors is proportional to the fundamental electronic correlation between the organic ligands and the Ln(III) ion in a Ln(III) complex $\left|\langle S_0, {}^5D_4|\boldsymbol{H}_{\mathrm{ET}}|T_1, {}^7F_6\rangle\right|$, and has been the central discussion of the ET between ligand and Ln(III) ions. The values indicate that fundamentally, the $T_1$ state of 4Me and 5Me ligand interacts strong electronically with the $^5D_4$ level of Tb(III) ions compared to that of Me ligands. The results also imply that the FET and BET processes can be expressed by potential energy surface as shown in Fig. 5.8. This argument on the existence of the activation energy in the ET process between ligands and Tb(III) ions is consistent with the previous report [14]. On the other hand, the nature of the other constants has been overlooked. In the particular case of these clusters, the energy gap is inconsistent with the lifetime results, but now can be explained by the activation energy and frequency factor. All of these three parameters, i.e., the energy gap $\Delta E$, the activation energy $Ea$, and the frequency factor were shown to be important in discussing BET in Ln(III) complexes.

## 5.4 Conclusion

A series of nonanuclear Tb(III) clusters with ester salicylate derivatives, $Tb_9$-Me, $Tb_9$-4Me, $Tb_9$-5Me, $Tb_9$-Et, $Tb_9$-Pr, $Tb_9$-Bu, and $Tb_9$-He has been synthesized and characterized by their photophysical properties. The difference in the photophysical

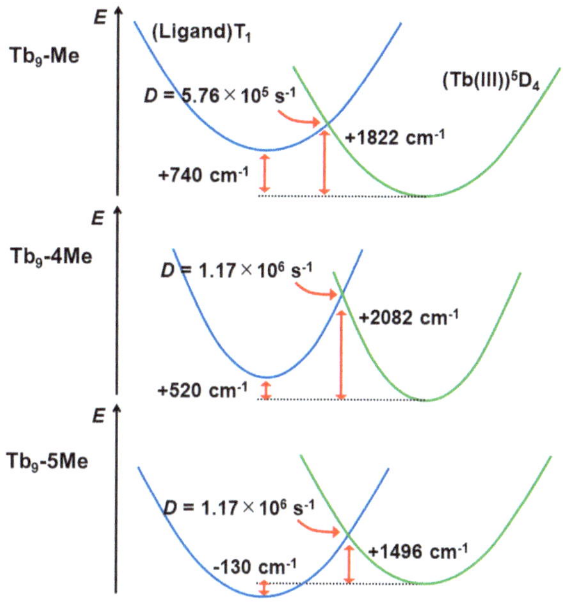


**Fig. 5.8** Schematic diagram of the potential energy curve of $T_1$ and $^5D_4$ states of the clusters

properties arising from the ligands depended on the length of the alkyl chain of the ligands. Meanwhile, the emission spectra of Tb(III) emission showed that the photophysical properties arising from the core structure of the cluster remained unaffected. The length of the alkyl chain of the ligands affects the CH–π and π–π interactions, which systematically changes the packing structure of the clusters, but does not affect the core structure of the cluster. It is suggested that the core structure of the cluster is rigid [15].

The photophysical properties of $Tb_9$-Me, $Tb_9$-4Me, and $Tb_9$-5Me, where the ligands have the same alkyl chain length but the different position of the methyl substituent on the aromatic right, were drastically different. $Tb_9$-4Me showed more than four times the increase in the lifetime compared to $Tb_9$-Me, and three times the increase compared to $Tb_9$-5Me. DFT calculations revealed that the position of the methyl group is directly linked to the HOMO and LUMO energy of the ligands. The introduction of the methyl group and its position on the aromatic ring also slightly altered the $S_0$–$S_1$ and $S_0$–$T_1$ energy gap. This affected the photophysical properties of the clusters significantly due to BET that is sensitive to the T1-5D4 energy gap $\Delta E$. Meanwhile, the energy gaps and the emission lifetimes were not consistent: the energy gap was the largest for $Tb_9$-Me, but $Tb_9$-4Me showed the longest lifetime. The plot of emission lifetime temperature dependence was fitted with Arrhenius equation and lead to the conclusion that the activation energy $Ea$ and the frequency factor $D$ of the energy transfer was the main contributor to the lifetimes, not the energy gap $\Delta E$, for these clusters. The results provide new insight into the BET and FET mechanism of Ln(III) complexes [16].

## 5.5 Appendix

### *5.5.1 Intrinsic Quantum Yield Measurements*

Absolute quantum yield measurement uses integration sphere. The measurement monitors the emission intensity of blank sample and the cluster sample. This requires that the cluster sample is emitting enough for adequate quantitative detection. As shown in Fig. 5.9, due to the low intensity of the emission observed for direct excitation of Tb(III) ions, the value obtained here, is still unreliable.

The intrinsic emission quantum yield $\Phi_{Ln}$ can also be calculated by Eq. (1.4):

$$\Phi_{Ln} = \frac{kr}{kr + knr} = \frac{\tau_{obs}}{\tau_{rad}}, \tag{1.4}$$

where $\tau_{rad} = 1/kr)$ is the emission lifetime of Ln(III) in the ideal state where there are no quenching. Radiative rate constant $kr$ is estimated as emission lifetime at 100 K under the assumption that lifetime at low temperature is purely radiative (non-radiative rate constant $knr$ is negligible) [17, 18]. This assumption is of course not very accurate and also does not yield a very reliable values. The results are summarized in Table 5.3.

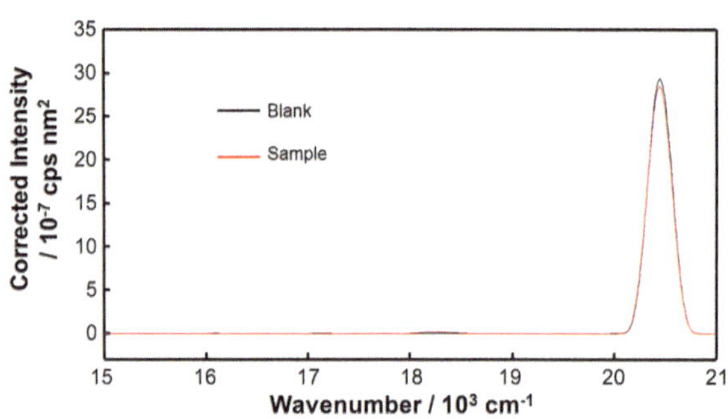


**Fig. 5.9** Emission spectra measured in integration sphere with corrections

**Table 5.3** Photophysical properties of $Tb_9$-Me, $Tb_9$-4Me, and $Tb_9$-5Me

| Cluster | $\Phi_{\pi\pi^*}$ (%) | $\Phi_{Ln}$ (%) | $\eta_{sens}$ (%) | $\tau_{obs}$/ms | $\tau_{rad}$/ms | $k_r/s^{-1}$ | $k_{nr}/s^{-1}$ |
|---|---|---|---|---|---|---|---|
| $Tb_9$-Me | 31 | 57 | 54 | 0.815 | 1.44 | 696 | 531 |
| $Tb_9$-4Me | 2.4 | 16 | 15 | 0.190 | 1.22 | 823 | 4430 |
| $Tb_9$-5Me | 6.7 | 17 | 39 | 0.264 | 1.53 | 654 | 3127 |

### 5.5.2 BET Rate Constant

BET can be considered as a transition from $^5D_4$ state to $T_1$ state coupled by Coulomb interaction between electrons. Taking the Coulomb potential as a perturbation in a system (the cluster), the transition rate constant can be expressed by Fermi's Golden Rule. Applying the Marcus-type equation to ET rate constant under the assumption that the BET is an adiabatic process, $k_{\text{BET}}$ can be expressed in the following equation [19–21]:

$$k_{\text{BET}} = \frac{2\pi}{\hbar}\left|\left\langle S_0, {}^5D_4 \middle| \boldsymbol{H}_{\text{ET}} \middle| T_1, {}^7F_6 \right\rangle\right|^2 \frac{1}{\sqrt{4\pi\lambda k_B T}} \exp\left(-\frac{Ea}{k_B T}\right), \tag{5.4}$$

where $\hbar$, $\left\langle S_0, {}^5D_4 \middle| \boldsymbol{H}_{\text{ET}} \middle| T_1, {}^7F_6 \right\rangle$, $\lambda$, $k_B$, $T$, $Ea$, are Planck constant, matrix element of the ET from $S_0$,$^5D_4$ to $T_1$,$^7F_6$ states (BET), reorganization energy, Boltzmann constant, temperature, and activation energy of BET. Reorganization energy $\lambda$ is given by:

$$\lambda = 2Ea\left(1 - \frac{\Delta E}{2Ea} + \sqrt{1 - \frac{\Delta E}{Ea}}\right). \tag{5.5}$$

By combining Eqs. (5.1) and (5.2), the following is derived:

$$k_{\text{BET}} = \frac{1}{\tau(T)} - \frac{1}{\tau(87\text{K})} = \frac{2\pi}{\hbar}\left|\left\langle S_0, {}^5D_4 \middle| \boldsymbol{H}_{\text{ET}} \middle| T_1, {}^7F_6 \right\rangle\right|^2 \frac{1}{\sqrt{4\pi\lambda k_B T}} \exp\left(-\frac{Ea}{k_B T}\right). \tag{5.6}$$

Equation (5.4) is often simplified to Eq. (5.2):

$$k_{\text{BET}} = D \exp\left(-\frac{Ea}{k_B T}\right), \tag{5.2}$$

where D becomes the frequency factor that includes matrix element of the BET. $\frac{1}{\sqrt{4\pi\lambda k_B T}}$ is also temperature dependent in Eq. (5.6) but will be neglected here. For example, $\frac{1}{\sqrt{4\pi\lambda k_B T}}$ changes by a factor of 0.48 when the temperature changes from 87 to 380 K. Such temperature changes can be ignored as long as the clusters are considered at a same temperature.

The Arrhenius plot using Eq. (5.2) is given in Fig. 5.10.

### 5.5.3 BET and $T_1$ Decay

As explained in Sect. 2.2.3, in the presence of BET, the emission lifetime is not precisely expressed simply by Eq. (5.1). Looking at Eq. (2.58) and under

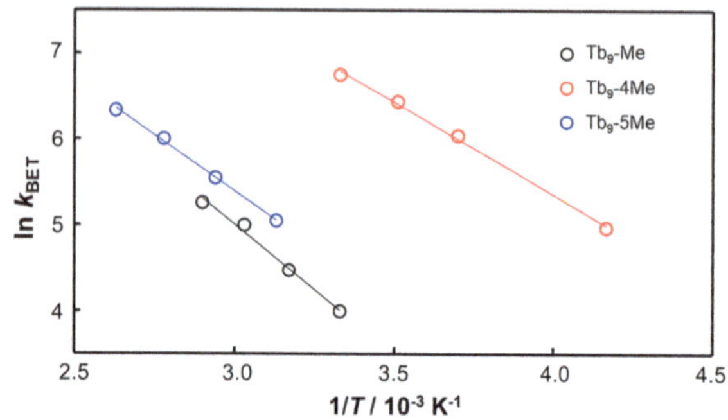


**Fig. 5.10** Arrhenius plots of the emission lifetimes temperature dependence of the clusters

the circumstance that in most cases, $S_1$ and $T_1$ states have very short lifetimes ($a_{11} \gg a_{22} \gg a_{33}$), the total relaxation rate constant for a Tb(III) ion $k_{Tb,actual}$ is expressed as:

$$k_{Tb,actual} = \frac{k_{T1} + k_{Tb} + k_{FET} + k_{BET} - \sqrt{(k_{T1} - k_{Tb})^2 + (k_{FET} + k_{BET})^2 + 2(k_{T1} - k_{Tb})(k_{FET} + k_{BET})}}{2}, \quad (5.7)$$

where $k_{T1}$ and $k_{Tb}$ are intrinsic relaxation rate of $T_1$ state and Ln excited state, respectively. Under the assumption that the $T_1$ state and the $^5D_4$ state of the clusters can be expressed with the potential energy diagrams shown in Fig. 5.8, $k_{FET} = k_{BET}\frac{1}{\exp(-\frac{\Delta E}{kT})}$ and the frequency factor can be assumed same for FET and BET near the intersection, and thus Eq. (5.7) converts to:

$$k_{Tb,actual} = \frac{k_{T1} + k_{Tb} + k_{BET}\left(\frac{1}{\exp\left(-\frac{\Delta E}{kT}\right)} + 1\right) - \sqrt{(k_{T1} - k_{Tb})^2 + k_{BET}^2\left(\frac{1}{\exp\left(-\frac{\Delta E}{kT}\right)} + 1\right)^2 + 2k_{BET}(k_{T1} - k_{Tb})\left(\frac{1}{\exp\left(-\frac{\Delta E}{kT}\right)} - 1\right)}}{2}. \quad (5.8)$$

Equation (5.8) implies the decay rate of $T_1$ state also affect the emission lifetimes of $^5D_4$ state (not just BET rate constant).

Meanwhile, this study concerns comparison between three $Tb_9$ clusters with different ligands. As long as the $k_{T1}$ is identical between the ligands, the comparison can still be restricted to the BET rate and the energy gap between $T_1$ and $^5D_4$ states. TD-DFT calculation of the $S_1$ state of Li-Me, Li-4Me, and Li-5Me resulted in an $S_1 \rightarrow S_0$ oscillator strengths of $f = 0.1377$, 0.1368, and 0.1363, respectively (corresponding to a pure HOMO-LUMO transition). Along with the close structure of the ligands, we can assume that the $k_{T1}$ are nearly identical between these three clusters. Therefore, the comparison can still be restricted to the BET rate constants $k_{BET}$ of the clusters.

# References


1. Smentek L, Kędziorski A (2010) Efficiency of the energy transfer in lanthanide-organic chelates; spectral overlap integral. J Lumin 130:1154–1159
2. Hasegawa Y, Wada Y, Yanagida S (2004) Strategies for the design of luminescent lanthanide (III) complexes and their photonic applications. J Photochem Photobiol, C 5:183–202
3. Gutierrez F, Tedeschi C, Maron L, Daudey J-P, Poteau R, Azema J, Tisnès P, Picard C (2004) Quantum chemistry-based interpretations on the lowest triplet state of luminescent lanthanides complexes. Part 1. Relation between the triplet state energy of hydroxamate complexes and their luminescence properties. Dalton Trans 1334–1347
4. Latva M, Takalo H, Mukkala V-M, Matachescu C, Rodríguez-Ubis JC, Kankare J (1997) Correlation between the lowest triplet state energy level of the ligand and lanthanide (III) luminescence quantum yield. J Lumin 75:149–169
5. Nakanishi T, Suzuki Y, Doi Y, Seki T, Koizumi H, Fushimi K, Fujita K, Hinatsu Y, Ito H, Tanaka K, Hasegawa Y (2014) Enhancement of optical faraday effect of nonanuclear Tb (III) complexes. Inorg Chem 53:7635–7641
6. Lima NBD, Gonçalves SMC, Júnior SA, Simas AM (2013) A comprehensive strategy to boost the quantum yield of luminescence of europium complexes Sci Rep 3, 2395
7. Miyata K, Nakagawa T, Kawakami R, Kita Y, Sugimoto K, Nakashima T, Harada T, Kawai T, Hasegawa Y (2011) Remarkable luminescence properties of lanthanide complexes with asymmetric dodecahedron structures. Chem Eur J 17:521–528
8. Yanagisawa K, Nakanishi T, Kitagawa Y, Seki T, Akama T, Kobayashi M, Taketsugu T, Ito H, Fushimi K, Hasegawa Y (2015) Seven-coordinate luminophores: brilliant luminescence of lanthanide complexes with C3v geometrical structures. Eur J Inorg Chem 4769–4774
9. Hasegawa Y, Ohkubo T, Nakanishi T, Kobayashi A, Kato M, Seki T, Ito H, Fushimi K (2013) Effect of ligand polarization on asymmetric structural formation for strongly luminescent lanthanide complexes. Eur J Inorg Chem 5911–5918
10. Manseki K, Hasegawa Y, Wada Y, Ichida H, Kanematsu Y, Kushida T (2007) Visible and near-infrared luminescence from self-assembled lanthanide (III) clusters with organic photosensitizers. J Lumin 122–123:262–264
11. Katagiri S, Tsukahara Y, Hasegawa Y, Wada Y (2007) Energy-transfer mechanism in photoluminescent terbium (III) complexes causing their temperature-dependence. Bull Chem Soc Jpn 80:1492–1503
12. Yanagisawa K, Kitagawa Y, Nakanishi T, Akama T, Kobayashi M, Seki T, Fushimi K, Ito H, Taketsugu T, Hasegawa Y (2017) Enhanced Luminescence of Asymmetrical Seven-Coordinate EuIII Complexes Including LMCT Perturbation. Eur J Inorg Chem 3843–3848
13. Miyata K, Konno Y, Nakanishi T, Kobayashi A, Kato M, Fushimi K, Hasegawa Y (2013) Chameleon luminophore for sensing temperatures: control of metal-to-metal and energy back transfer in lanthanide coordination polymers. Angew Chem Int Ed 52:6413–6416
14. Hatanaka M, Morokuma K (2014) The ONIOM method and its applications. J Chem Theory Comput 10:4184–4188
15. Omagari S, Nakanishi T, Kitagawa Y, Fushimi K, Hasegawa Y (2015) Synthesis and photoluminescence properties of nonanuclear Tb (III) clusters with long alkyl chain group. e-J Surf Sci Nanotech 13, 27–30
16. Omagari S, Nakanishi T, Seki T, Kitagawa Y, Takahata Y, Fushimi K, Ito H, Hasegawa Y (2015) effective photosensitized energy transfer of nonanuclear terbium clusters using methyl salicylate derivatives. J Phys Chem A 119, 1943–1947
17. Nasso I, Geum N, Bechara G, Mestre-Voegtlé B, Galaup C, Picard C (2014) Highly luminescent Tb (III) macrocyclic complex based on a DO3A hosting unit and an appended carboxylated N, C-pyrazolylpyridine antenna. J Photochem Photobiol, A 274:124–132
18. Bassett AP, Magennis SW, Glover PB, Lewis DJ, Spencer N, Parsons S, Williams RM, De Cola L, Pikramenou Z (2004) Highly luminescent, triple-and quadruple-stranded, dinuclear Eu, Nd, and Sm (III) lanthanide complexes based on bis-diketonate ligands. J Am Chem Soc 126:9413–9424

19. Marcus RA, Sutin N (1985) Electron transfers in chemistry and biology. Biochim Biophys Acta 811:265–322
20. Marcus RA (1984) Nonadiabatic processes involving quantum-like and classical-like coordinates with applications to nonadiabatic electron transfers. J Chem Phys 81:4494–4500
21. Sudha Devi L, Al-Suti MK, Dosche C, Khan MS, Friend RH, Köhler A (2008) Triplet energy transfer in conjugated polymers. I. Experimental investigation of a weakly disordered compound Phys Rev B 78, 1–8

# Chapter 6
# Suppression of Back Energy Transfer by Energy Transfer Between Terbium Ions

**Abstract** Suppression of back energy transfer is crucial in realizing efficient luminescent lanthanide complexes. However, the only practical method reported to this day is to raise the energy of the triplet excited state so that back energy transfer is energetically unfavorable, which limits the application where the absorbing wavelength of the organic ligands is important. This chapter explores a radically new strategy that focus on the yield of the back energy transfer as opposed to the rate constant of the back energy transfer. By utilizing energy transfer between lanthanide ions, which is a competitive process to back energy transfer, the contribution of back energy transfer can be suppressed. This chapter describes the theoretical background of this concept using kinetic analysis, and then the experimental confirmation of the proposal using $[Tb_nGd_{9-n}(\mu\text{-}OH)_{10}(Bu)_{16}]NO_3$ ($n = 0, 1, 2, 5, 8, 9$). It is revealed that indeed the contribution of the back energy transfer is suppressed in this cluster, implying the potential of lanthanide clusters as functionalized and efficient luminescent material.

## 6.1 Introduction

So far, various factors that affect the ET processes in Ln(III) complexes have been discussed. In Chap. 3, the effect of YbYbET on the photophysical properties of Yb(III) coordination polymers was elucidated. Chapter 4 described the effect of SOC on the sensitization efficiency by using nonanuclear Yb(III)/Gd(III) and Yb(III)/Lu(III) clusters. The effect of BET and its mechanism in Chap. 5 were explored using nonanuclear Tb(III) clusters. These chapters provided new insights into LnLnET of polynuclear Ln(III) complexes as well as some fundamental properties of nonanuclear Ln(III) clusters and factors that affect sensitization efficiency. Now it is set to combine these insights and focus on LnLnET in nonanuclear Ln(III) clusters and its effect on the photophysical properties.

The hypothesis here is that the ET between Tb(III) ions (TbTbET) can suppress BET in Tb(III) complexes. Considering that both BET and TbTbET occur from the same 4f-excited state of Tb(III) ion ($^5D_4$), these two processes are competitive. When TbTbET is adequately larger than BET, the overall contribution of BET could be suppressed. In this regard, nonanuclear Tb(III) clusters are a good candidate since the



S. Omagari, *Energy Transfer Processes in Polynuclear Lanthanide Complexes*, Springer Theses, https://doi.org/10.1007/978-981-13-6049-7_6

Ln-Ln distance is quite short at around 3.65 Å. The structural feature of nonanuclear Tb(III) cluster is also important. The center Tb(III) ion is coordinated only with OH- ions (see Sect. 4.3.1) and thus the center Tb(III) ion is expected not to involve (or involve negligibly little) in FET and BET compared to those of the other Tb(III) ions (Fig. 6.1). Therefore, TbTbET to this center Tb(III) ion overall raises the probability of the cluster emitting from the Tb(III) ions than BET to the ligands. The details of this concept are discussed in Sect. 6.2 of this chapter (theoretical motivation). This new strategy implies that BET can be suppressed without changing $T_1$ state energy, which may become important for applications such as spectral converters where both the absorption and emission wavelength are specified [1–3]. For example, two-color white LED consists of blue GaN-based LED chip with blue absorbing, yellow emitting Ce:YAG spectral converter. Fluorescent lamps use spectral converters that absorb in the UV region and emit in blue, green, and red [4].

In this study, Gd(III) ions were introduced to nonanuclear Tb(III) clusters to investigate the relationship between TbTbET and BET. Gd(III) ions were used as spacers between Tb(III) ions to modulate the TbTbET processes. Novel nonanuclear Ln(III) clusters containing Tb(III) and Gd(III) ions $[Tb_nGd_{9\text{-}n}(\mu\text{-OH})_{10}(Bu)_{16}]NO_3$ ($n = 0, 1, 2, 5, 8, 9$) (referred to as $Tb_nGd_{9-n}$-Bu) were synthesized by complexation reaction of Tb(III)/Gd(III) nitrate salt and Bu ligands. Their characteristic structures were evaluated using mass spectroscopy and powder XRD. Emission spectroscopy, quantum yield, and emission lifetime measurements were performed for the clusters in a chloroform solution. Temperature dependence of emission lifetimes was also measured for the estimation of the activation energy of BET. The experimental results are compared with that of theoretical motivation (Sect. 6.2) to qualitatively see the suppression effect. The promising properties of Ln(III) clusters and their potential as a new class of highly efficient luminescent materials are presented in this chapter.

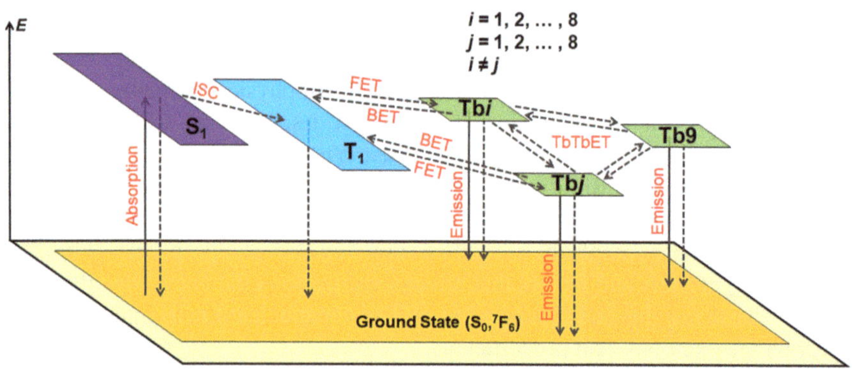


**Fig. 6.1** Schematic 3D energy diagram depicting the ET processes in nonanuclear Tb(III) clusters

## 6.2 Theoretical Motivation

### 6.2.1 *Theoretical Method*

The importance of TbTbET and a Tb(III) ion site unaffected by BET and FET in suppressing BET has been briefly mentioned in Sect. 6.1. This section elaborates on this hypothesis by taking nonanuclear Tb(III) cluster as a model and qualitatively calculate the theoretical emission lifetimes and BET efficiencies by rate equation analysis [5].

First, the Tb(III) ions are labeled as "Tb1", "Tb2", ..., and "Tb9" (where 9 is the center Tb(III) ion) as shown in Fig. 6.2 to distinguish between Tb(III) ions of a different site. The dynamics model of the energy transfer processes in $Tb_9$-Bu is shown in Fig. 6.3. Three assumptions are made here that does not affect the conclusion of the section in any way but are important in the calculations for the sake of discussion and simplicity. (1) Both FET and BET are ignored for Tb9 (the center Tb(III) ion). Tb9 is not directly coordinated with the Bu ligands. Due to the strong distance sensitivity of ET, the ET between Bu ligands and Tb9 is significantly more inefficient compared to the other Tb(III) ions that are coordinated directly with the Bu ligands ($m = 1, 2, \ldots, 8$). This means that Tb9 can only be populated by TbTbET. (2) TbTbET is calculated based on the $R^{-6}$ distance dependence (dipole-dipole interaction), and all the other interactions are neglected (i.e., exchange, dipole-quadrupole, and quadrupole-quadrupole interactions) [6, 7]. While the Tb–Tb distance of 3.65 Å is close enough for all types of interactions, considering these interactions would only enhance TbTbET, thereby enhancing the BET suppression effect. Thus, for this qualitative investigation, calculating the dipole-dipole interaction is adequate. (3) The rate constants used for the calculations are based on the previous reports on different systems. The rate constants are arbitrary (as long as they are not zero), but to keep the rate constant within a reasonable scale, values previously reported were used.

The excited state dynamics can be expressed as a system of linear differential equations in the matrix form as introduced in Sect. 2.3:

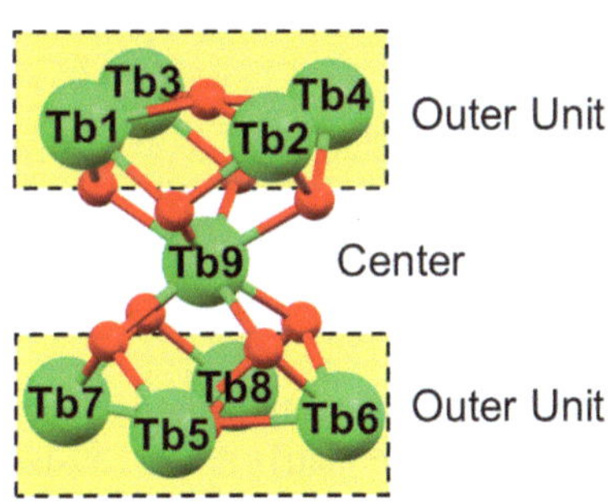


**Fig. 6.2** Labeling of the Tb(III) ions in $Tb_9$-Bu. Reproduced from Ref. [21] with permission from Springer Nature

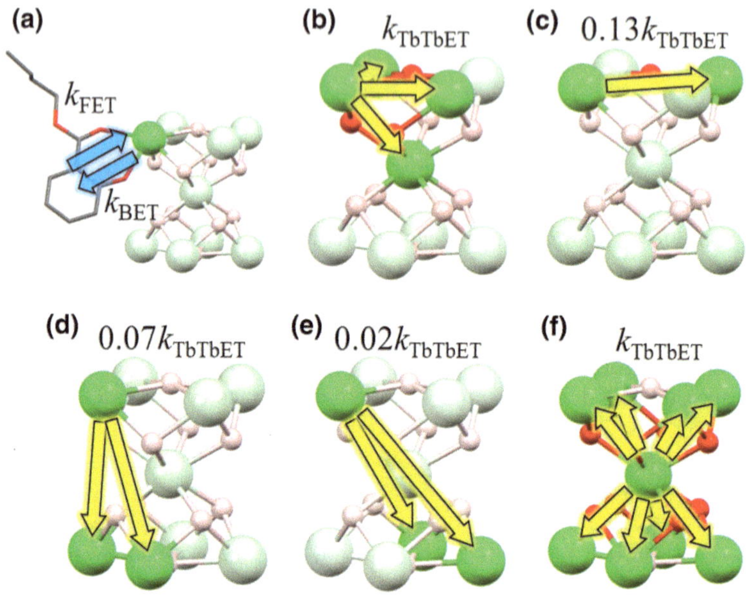


**Fig. 6.3** Excited state dynamics model of $Tb_9$-Bu. **a** Between a Tb(III) ion and ligands, **b** closest Tb(III) ion pair of the same outer unit, **c** Tb(III) ion pair on the opposite of the same outer unit, **d** closest Tb(III) ion pair of different outer unit, **e** Tb(III) ion pair on the opposite of the different outer unit, and **f** Tb9 and all the other Tb(III) ions. Reproduced from Ref. [21] with permission from Springer Nature

$$\frac{d\boldsymbol{X}(t)}{dt} = \boldsymbol{A}\boldsymbol{X}(t) + \boldsymbol{J}(t), \tag{2.36}$$

where $\boldsymbol{X}(t)$ is the vector of population density of a specific state. In this case, there are 11 excited states, namely singlet excited state $\mathrm{S1}(t)$, triplet excited state $\mathrm{T1}(t)$, excited Tb(III) ions in the outer unit $\mathrm{Tb}m(t)$ ($m = 1, 2, \ldots, 8$), and the center Tb(III) ion ($\mathrm{Tb9}(t)$):

$$\boldsymbol{X}(t) = \begin{pmatrix} \mathrm{S1}(t) \\ \mathrm{T1}(t) \\ \mathrm{Tb1}(t) \\ \vdots \\ \mathrm{Tb8}(t) \\ \mathrm{Tb9}(t) \end{pmatrix}. \tag{6.1}$$

A is the "interaction matrix" that models the system. With the assumptions above, the interaction matrix A for Tb9-Bu becomes:

$$
A = \begin{bmatrix}
-\frac{1}{\tau_{\mathrm{S}1}} & 0 & 0 & 0 & 0 & 0 & 0 & 0 & 0 & 0 & 0 \\
k_{\mathrm{isc}} & -\frac{1}{\tau_{\mathrm{T}1}} & k_{\mathrm{BET}} & k_{\mathrm{BET}} & k_{\mathrm{BET}} & k_{\mathrm{BET}} & k_{\mathrm{BET}} & k_{\mathrm{BET}} & k_{\mathrm{BET}} & k_{\mathrm{BET}} & 0 \\
0 & k_{\mathrm{FET}} & -\frac{1}{\tau_{\mathrm{Tb}m}} & k_{\mathrm{TbTbET}} & k_{\mathrm{TbTbET}} & 0.13k_{\mathrm{TbTbET}} & 0.07k_{\mathrm{TbTbET}} & 0.02k_{\mathrm{TbTbET}} & 0.07k_{\mathrm{TbTbET}} & 0.02k_{\mathrm{TbTbET}} & k_{\mathrm{TbTbET}} \\
0 & k_{\mathrm{FET}} & k_{\mathrm{TbTbET}} & -\frac{1}{\tau_{\mathrm{Tb}m}} & 0.13k_{\mathrm{TbTbET}} & k_{\mathrm{TbTbET}} & 0.07k_{\mathrm{TbTbET}} & 0.07k_{\mathrm{TbTbET}} & 0.02k_{\mathrm{TbTbET}} & 0.02k_{\mathrm{TbTbET}} & k_{\mathrm{TbTbET}} \\
0 & k_{\mathrm{FET}} & k_{\mathrm{TbTbET}} & 0.13k_{\mathrm{TbTbET}} & -\frac{1}{\tau_{\mathrm{Tb}m}} & k_{\mathrm{TbTbET}} & 0.02k_{\mathrm{TbTbET}} & 0.07k_{\mathrm{TbTbET}} & 0.02k_{\mathrm{TbTbET}} & 0.07k_{\mathrm{TbTbET}} & k_{\mathrm{TbTbET}} \\
0 & k_{\mathrm{FET}} & 0.13k_{\mathrm{TbTbET}} & k_{\mathrm{TbTbET}} & k_{\mathrm{TbTbET}} & -\frac{1}{\tau_{\mathrm{Tb}m}} & 0.02k_{\mathrm{TbTbET}} & 0.02k_{\mathrm{TbTbET}} & 0.07k_{\mathrm{TbTbET}} & 0.07k_{\mathrm{TbTbET}} & k_{\mathrm{TbTbET}} \\
0 & k_{\mathrm{FET}} & 0.07k_{\mathrm{TbTbET}} & 0.07k_{\mathrm{TbTbET}} & 0.02k_{\mathrm{TbTbET}} & 0.02k_{\mathrm{TbTbET}} & -\frac{1}{\tau_{\mathrm{Tb}m}} & k_{\mathrm{TbTbET}} & k_{\mathrm{TbTbET}} & 0.13k_{\mathrm{TbTbET}} & k_{\mathrm{TbTbET}} \\
0 & k_{\mathrm{FET}} & 0.02k_{\mathrm{TbTbET}} & 0.07k_{\mathrm{TbTbET}} & 0.07k_{\mathrm{TbTbET}} & 0.02k_{\mathrm{TbTbET}} & k_{\mathrm{TbTbET}} & -\frac{1}{\tau_{\mathrm{Tb}m}} & 0.13k_{\mathrm{TbTbET}} & k_{\mathrm{TbTbET}} & k_{\mathrm{TbTbET}} \\
0 & k_{\mathrm{FET}} & 0.07k_{\mathrm{TbTbET}} & 0.02k_{\mathrm{TbTbET}} & 0.02k_{\mathrm{TbTbET}} & 0.07k_{\mathrm{TbTbET}} & k_{\mathrm{TbTbET}} & 0.13k_{\mathrm{TbTbET}} & -\frac{1}{\tau_{\mathrm{Tb}m}} & k_{\mathrm{TbTbET}} & k_{\mathrm{TbTbET}} \\
0 & k_{\mathrm{FET}} & 0.02k_{\mathrm{TbTbET}} & 0.02k_{\mathrm{TbTbET}} & 0.07k_{\mathrm{TbTbET}} & 0.07k_{\mathrm{TbTbET}} & 0.13k_{\mathrm{TbTbET}} & k_{\mathrm{TbTbET}} & k_{\mathrm{TbTbET}} & -\frac{1}{\tau_{\mathrm{Tb}m}} & k_{\mathrm{TbTbET}} \\
0 & 0 & k_{\mathrm{TbTbET}} & k_{\mathrm{TbTbET}} & k_{\mathrm{TbTbET}} & k_{\mathrm{TbTbET}} & k_{\mathrm{TbTbET}} & k_{\mathrm{TbTbET}} & k_{\mathrm{TbTbET}} & k_{\mathrm{TbTbET}} & -\frac{1}{\tau_{\mathrm{Tb}9}}
\end{bmatrix}, \tag{6.2}
$$

where

$$\frac{1}{\tau_{S1}} = kr_{S1} + knr_{S1} + k_{isc}, \tag{6.3}$$

$$\frac{1}{\tau_{T1}} = kr_{T1} + knr_{T1} + 8k_{FET}, \tag{6.4}$$

$$\frac{1}{\tau_{Tbm}} = kr_{Tb} + knr_{Tb} + 3.31k_{TbTbET} + k_{BET}, \tag{6.5}$$

$$\frac{1}{\tau_{Tb9}} = kr_{Tb} + knr_{Tb} + 8k_{TbTbET}, \tag{6.6}$$

$k_{isc}$, $kr$, $knr$, $k_{FET}$, $k_{BET}$, and $k_{TbTbET}$ are defined as intersystem-crossing, radiative, nonradiative, FET, BET, and TbTbET rate constants, respectively. The subscript indicates a species in $Tb_9$-Bu. Finally, vector $\boldsymbol{J}(t)$ represents the input function of each species in $Tb_9$-Bu:

$$\boldsymbol{J}(t) = \begin{pmatrix} J_{S1}(t) \\ J_{T1}(t) \\ J_{Tb1}(t) \\ \vdots \\ J_{Tb8}(t) \\ J_{Tb9}(t) \end{pmatrix}. \tag{6.7}$$

In $Tb_9$-Bu, the ligands are excited to the singlet excited state $S_1$, and all the other states are not directly excited by outside sources. Here, doubly excited states were ignored since excitation with reasonably low intensity (under 20 W $mm^{-2}$) is considered [8]. Thus, $J_{T1}(t) = J_{Tb1}(t) = \cdots = J_{Tb8}(t) = J_{Tb9}(t) = 0$. $J_{S1}(t)$ depends on the purpose of the calculation. If the emission lifetimes are to be calculated, then laser pulse excitation is assumed and $J_{S1}(t)$ is expressed by an impulse function $\delta(t)$ (Dirac's delta function):

$$\boldsymbol{J}(t) = \begin{pmatrix} \delta(t)j_1 \\ 0 \\ 0 \\ \vdots \\ 0 \\ 0 \end{pmatrix} = j_1 \begin{pmatrix} \delta(t) \\ 0 \\ 0 \\ \vdots \\ 0 \\ 0 \end{pmatrix}, \tag{6.8}$$

where the constant $j_1$ is the amplitude of the excitation (the intensity of the excitation source and the absorption of the $S_0 \rightarrow S_1$ transition). Since the intensity in terms of lifetimes are arbitrary, constant $j_1$ can be any value other than zero. For simplicity, $j_1 = 1$, which means that the excitation intensity is normalized. Calculation under this input function will lead to time-dependent population density, which is linearly proportional to emission lifetimes. If the quantum yields are to be calculated, then we are assuming steady excitation and $J_{S1}(t)$ is expressed:

$$\boldsymbol{J}(t) = \begin{pmatrix} j_1 \\ 0 \\ 0 \\ \vdots \\ 0 \\ 0 \end{pmatrix} = j_1 \begin{pmatrix} 1 \\ 0 \\ 0 \\ \vdots \\ 0 \\ 0 \end{pmatrix}, \qquad (6.9)$$

where the constant $j_1$, once again, is the amplitude of the excitation. If the excitation intensity is normalized ($j_1 = 1$), then the efficiency of a process can be calculated. Here, the focus is the emission quantum yield $\Phi_{\pi\pi^*,\mathrm{calc}}$ and the BET efficiency $\eta_{\mathrm{BET,calc}}$. The BET efficiency $\eta_{\mathrm{BET,calc}}$ is defined as the percentage of loss of population density of Tb(III) ions in the presence of BET to when BET is absent:

$$\eta_{\mathrm{BET,calc}} = \frac{[\mathrm{Tb1}(t) + \mathrm{Tb2}(t) + \cdots + \mathrm{Tb9}(t)]_{k_{\mathrm{BET}}=0} - [\mathrm{Tb1}(t) + \mathrm{Tb2}(t) + \cdots + \mathrm{Tb9}(t)]_{k_{\mathrm{BET}}\neq 0}}{[\mathrm{Tb1}(t) + \mathrm{Tb2}(t) + \cdots + \mathrm{Tb9}(t)]_{k_{\mathrm{BET}}=0}}. \qquad (6.10)$$

The rate constants used for this calculation are summarized in Table 6.1 [9, 10]. The values of $k_{\mathrm{FET}}$ and $k_{\mathrm{BET}}$ are arbitrary but 6000 and 3000 $s^{-1}$, respectively, were for clarity. Here, two cases are compared: (1) effect of TbTbET in the absence of BET, and (2) effect of TbTbET in the presence of BET on the photophysical properties of $Tb_9$-Bu. The calculations were performed on MATLAB, and the scripts (algorithm) are provided in Sect. 6.6.2.

### 6.2.2 Results

Figure 6.4 shows the calculated population density decay profile of $Tb_9$-Bu after an impulse excitation. In the absence of BET (Fig. 6.4a), the decay profile for $k_{\mathrm{TbTbET}} = 0\ s^{-1}$ (absence of TbTbET: blue line) and $k_{\mathrm{TbTbET}} = 50{,}000\ s^{-1}$ (presence of TbTbET: red line) completely overlaps. The lifetime $\tau_{\mathrm{calc}}$ was calculated to be 1.30 ms from the slope. The emission quantum yield $\Phi_{\pi\pi^*,\mathrm{calc}}$ was also the same value of 25.3%. Calculating for $k_{\mathrm{TbTbET}} = 0$, 500, 5000, and 50,000 $s^{-1}$ still produced the same result (see Table 6.6). These results indicate that the TbTbET does not affect the

**Table 6.1** Rate constants used for the calculations

| Species | $k_r + k_{nr}/s^{-1}$ | $k_{isc}/s^{-1}$ | $k_{FET}/s^{-1}$ | $k_{BET}/s^{-1}$ |
|---|---|---|---|---|
| S1 | $3.4 \times 10^8$ [a] | $3.4 \times 10^8$ [a] | 0 | 0 |
| T1 | $1.5 \times 10^4$ [a] | 0 | $6.0 \times 10^3$ | 0 |
| Tb1–Tb8 | $8.0 \times 10^2$ [b] | 0 | 0 | $3.0 \times 10^3$ or 0 |
| Tb9 | $8.0 \times 10^2$ [b] | 0 | 0 | 0 |

[a]From Ref. [9]
[b]From Ref. [10]

observable photophysical properties of $Tb_9$-Bu. Meanwhile, in the presence of BET (Fig. 6.4b), the decay profile for $k_{TbTbET} = 50{,}000\ s^{-1}$ is longer compared to the case for $k_{TbTbET} = 0\ s^{-1}$. The lifetime was extended from $\tau_{calc} = 685$–$720\ \mu s$. The emission quantum yield $\Phi_{\pi\pi^*,calc}$ increased from 13.4 to 14.1% and the BET efficiency $\eta_{BET,calc}$ decreased from 47.2 to 44.3%. Furthermore, the population density of $T_1$ in the decay profile (dashed line in Fig. 6.4) was lower when $k_{TbTbET} = 50{,}000\ s^{-1}$. These results are summarized in Table 6.2. These results indicate that TbTbET can suppress BET in $Tb_9$-Bu.

To confirm the importance of Tb9 in $Tb_9$-Bu, a hypothetical cluster where Tb9 also involves in FET and BET the same way with other Tb(III) ions was also considered. The interaction matrix $\boldsymbol{A}'$ for this cluster is:

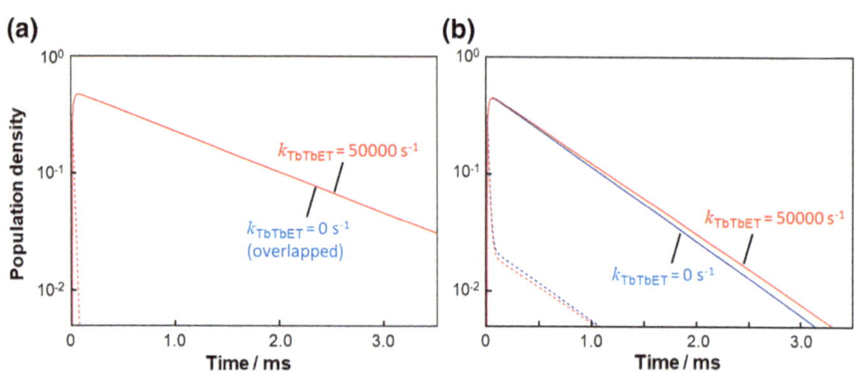


**Fig. 6.4** Calculated population density decay profiles **a** in the absence of BET, and **b** in the presence of BET. Solid lines: excited Tb(III) ions, dashed lines: the T1 state. Blue lines: absence of FET, red lines: presence of FET. Reproduced from Ref. [21] with permission from Springer Nature

$$
\boldsymbol{A}' = \begin{bmatrix}
-\frac{1}{\tau_{\mathrm{S1}}} & 0 & 0 & 0 & 0 & 0 & 0 & 0 & 0 & 0 & 0 \\
k_{\mathrm{isc}} & -\frac{1}{\tau_{\mathrm{T1}}} & k_{\mathrm{BET}} & k_{\mathrm{BET}} & k_{\mathrm{BET}} & k_{\mathrm{BET}} & k_{\mathrm{BET}} & k_{\mathrm{BET}} & k_{\mathrm{BET}} & k_{\mathrm{BET}} & k_{\mathrm{BET}} \\
0 & k_{\mathrm{FET}} & -\frac{1}{\tau_{\mathrm{Tb}m}} & k_{\mathrm{TbTbET}} & k_{\mathrm{TbTbET}} & 0.13k_{\mathrm{TbTbET}} & 0.07k_{\mathrm{TbTbET}} & 0.02k_{\mathrm{TbTbET}} & 0.07k_{\mathrm{TbTbET}} & 0.02k_{\mathrm{TbTbET}} & k_{\mathrm{TbTbET}} \\
0 & k_{\mathrm{FET}} & k_{\mathrm{TbTbET}} & -\frac{1}{\tau_{\mathrm{Tb}m}} & 0.13k_{\mathrm{TbTbET}} & k_{\mathrm{TbTbET}} & 0.07k_{\mathrm{TbTbET}} & 0.07k_{\mathrm{TbTbET}} & 0.02k_{\mathrm{TbTbET}} & 0.02k_{\mathrm{TbTbET}} & k_{\mathrm{TbTbET}} \\
0 & k_{\mathrm{FET}} & k_{\mathrm{TbTbET}} & 0.13k_{\mathrm{TbTbET}} & -\frac{1}{\tau_{\mathrm{Tb}m}} & k_{\mathrm{TbTbET}} & 0.02k_{\mathrm{TbTbET}} & 0.07k_{\mathrm{TbTbET}} & 0.02k_{\mathrm{TbTbET}} & 0.07k_{\mathrm{TbTbET}} & k_{\mathrm{TbTbET}} \\
0 & k_{\mathrm{FET}} & 0.13k_{\mathrm{TbTbET}} & k_{\mathrm{TbTbET}} & k_{\mathrm{TbTbET}} & -\frac{1}{\tau_{\mathrm{Tb}m}} & 0.02k_{\mathrm{TbTbET}} & 0.02k_{\mathrm{TbTbET}} & 0.07k_{\mathrm{TbTbET}} & 0.07k_{\mathrm{TbTbET}} & k_{\mathrm{TbTbET}} \\
0 & k_{\mathrm{FET}} & 0.07k_{\mathrm{TbTbET}} & 0.07k_{\mathrm{TbTbET}} & 0.02k_{\mathrm{TbTbET}} & 0.02k_{\mathrm{TbTbET}} & -\frac{1}{\tau_{\mathrm{Tb}m}} & k_{\mathrm{TbTbET}} & k_{\mathrm{TbTbET}} & 0.13k_{\mathrm{TbTbET}} & k_{\mathrm{TbTbET}} \\
0 & k_{\mathrm{FET}} & 0.02k_{\mathrm{TbTbET}} & 0.07k_{\mathrm{TbTbET}} & 0.07k_{\mathrm{TbTbET}} & 0.02k_{\mathrm{TbTbET}} & k_{\mathrm{TbTbET}} & -\frac{1}{\tau_{\mathrm{Tb}m}} & 0.13k_{\mathrm{TbTbET}} & k_{\mathrm{TbTbET}} & k_{\mathrm{TbTbET}} \\
0 & k_{\mathrm{FET}} & 0.07k_{\mathrm{TbTbET}} & 0.02k_{\mathrm{TbTbET}} & 0.02k_{\mathrm{TbTbET}} & 0.07k_{\mathrm{TbTbET}} & k_{\mathrm{TbTbET}} & 0.13k_{\mathrm{TbTbET}} & -\frac{1}{\tau_{\mathrm{Tb}m}} & k_{\mathrm{TbTbET}} & k_{\mathrm{TbTbET}} \\
0 & k_{\mathrm{FET}} & 0.02k_{\mathrm{TbTbET}} & 0.02k_{\mathrm{TbTbET}} & 0.07k_{\mathrm{TbTbET}} & 0.07k_{\mathrm{TbTbET}} & 0.13k_{\mathrm{TbTbET}} & k_{\mathrm{TbTbET}} & k_{\mathrm{TbTbET}} & -\frac{1}{\tau_{\mathrm{Tb}m}} & k_{\mathrm{TbTbET}} \\
0 & k_{\mathrm{FET}} & k_{\mathrm{TbTbET}} & k_{\mathrm{TbTbET}} & k_{\mathrm{TbTbET}} & k_{\mathrm{TbTbET}} & k_{\mathrm{TbTbET}} & k_{\mathrm{TbTbET}} & k_{\mathrm{TbTbET}} & k_{\mathrm{TbTbET}} & -\frac{1}{\tau_{\mathrm{Tb}m}}
\end{bmatrix}. \tag{6.11}
$$

**Table 6.2** Calculated lifetimes, quantum yields and BET efficiencies of $Tb_9$ cluster in the presence of BET

| $k_{TbET}/s^{-1}$ | $\tau_{calc}/\mu s$ | $\Phi_{\pi\pi*,calc}$ (%) | $\eta_{BET,calc}$ (%) |
|---|---|---|---|
| 50,000 (presence of TbET) | 720 | 14.1 | 44.3 |
| 0 (absence of TbET) | 685 | 13.4 | 47.2 |

Reproduced from Ref. [21] with permission from Springer Nature

**Table 6.3** Calculated quantum yield and BET efficiency of the hypothetical cluster with nine Tb(III) ions in direct contact with ligands

| $k_{TbET}/s^{-1}$ | $\tau_{calc}/\mu s$ | $\Phi_{\pi\pi*,calc}$ (%) | $\eta_{BET,calc}$ (%) |
|---|---|---|---|
| 50,000 (presence of TbET) | 713 | 14.31 | 44.90 |
| 0 (absence of TbET) | 713 | 14.31 | 44.90 |

The difference between Eqs. (6.2) and (6.11) is the 11th row and the 11th column, which express the interaction of Tb9 with the others. The elements (11, 2) and (2, 11) contain $k_{FET}$ and $k_{BET}$, respectively, for Eq. (6.11). In the case of Eq. (6.2), these elements are zero. The element (11, 11) in Eq. (6.11) is $-\frac{1}{\tau_{Tbm}}$ instead of $-\frac{1}{\tau_{Tb9}}$ as in Eq. (6.2). The same calculation as done above was once again performed.

The results of the calculation of the lifetime, emission quantum yield, and BET efficiency of the hypothetical cluster are summarized in Table 6.3. No difference in the values was observed in the presence and the absence of TbTbET. Above results unambiguously indicate that the energy migration between Tb(III) ions and a Tb(III) ion site unaffected by BET is crucial in suppressing BET.

In Sect. 6.4.2, the temperature dependences of emission lifetimes of $Tb_9$-Bu and $Tb_1Gd_8$-Bu are compared. $Tb_1Gd_8$-Bu does not show any energy transfer between Tb(III) ions. As mentioned in Chap. 5, BET proceeds provided that there is enough thermal energy to compensate for the energy difference between the excited state of Tb(III) ions and Bu ligands. Complete suppression of BET can thus be achieved at low temperature. Therefore, the investigation of the lifetimes as done in this section can be performed experimentally by comparing the emission lifetimes of $Tb_9$-Bu and $Tb_1Gd_8$-Bu at room and low temperatures.

## 6.3 Experimental Section

### 6.3.1 Material

Terbium(III) nitrate hexahydrate ($Tb(NO_3)_3{\cdot}6H_2O$) and triethylamine ($C_6H_{15}N$) were purchased from Kanto Chemical Co. Inc. Gadolinium(III) nitrate hexahydrate ($Gd(NO_3)_3{\cdot}6H_2O$) was purchased from Sigma-Aldrich Co. Butylsalicylate ($C_{11}H_{14}O_3$) was purchased from Tokyo Chemical Industry Co., LTD. All other chemicals and solvents were of reagent grade and were used without further purification.

### 6.3.2 *Apparatus*

FAB-MS spectra were measured on a JEOL JMS-700TZ. Elemental analyses were performed by Exter Analytical CE440. Infrared spectra were recorded on a JASCO FT/IR-4600 spectrometer. XRD spectra were characterized by a RIGAKU X-ray diffractometer RINT 2200. Single crystal X-ray diffractions were made on a RIGAKU RAXIS RAPID imaging plate area detector.

### 6.3.3 *Synthesis*

**General synthetic procedures of $Tb_nGd_{9-n}$-Bu**: Butylsalicylate (Bu, 1.05 g, 5.40 mmol) was dissolved in methanol, and triethylamine (1.22 mL, 8.80 mmol) was added to the solution to be stirred at 40 °C. A stoichiometric mixture of $Tb(NO_3)_3{\cdot}6H_2O$ and $Gd(NO_3)_3{\cdot}6H_2O$ (total of 3.04 mmol) in methanol was added dropwise to the solution. The solution stirred for 30 min and then cooled to room temperature. A white precipitate (target cluster) formed in the process. The precipitate was vacuum filtered and dried in vacuum.

**$Gd_9$-Bu ([$Gd_9(\mu$-OH$)_{10}$(Bu)16]$NO_3$)**: Yield: 59%. Selected IR (ATR, $cm^{-1}$): 3573 (w, O–H), 3234 (w, O–H), 2956 (m, C–H), 1318 (s, C–O). FAB-MS: m/z = 4676.8 [$Gd_9(\mu$-OH$)_{10}$(Bu$)_{16}$]$^+$. Elemental analysis calculated for $C_{176}H_{218}NO_{61}Gd_9$: C, 44.61%, H, 4.64%, N, 0.30%. Found: C, 44.30%, H, 4.53%, N, 0.25%.

**$Tb_1Gd_8$-Bu ([$Tb_1Gd_8(\mu$-OH$)_{10}$(Bu$)_{16}$]$NO_3$)**: Yield: 82%. Selected IR (ATR, $cm^{-1}$): 3573 (w, O–H), 3234 (w, O–H), 2956 (m, C–H), 1318 (s, C–O). FAB-MS: m/z = 4677.8 [$Tb_1Gd_8(\mu$-OH$)_{10}$(Bu$)_{16}$]$^+$. Elemental analysis calculated for $C_{176}H_{218}NO_{61}Tb_1Gd_8$: C, 44.59%, H, 4.64%, N, 0.30%. Found: C, 43.95%, H, 4.53%, N, 0.29%.

**$Tb_2Gd_7$-Bu ([$Tb_2Gd_7(\mu$-OH$)_{10}$(Bu$)_{16}$]$NO_3$)**: Yield: 51%. Selected IR (ATR, $cm^{-1}$): 3573 (w, O–H), 3234 (w, O–H), 2956 (m, C–H), 1318 (s, C–O). FAB-MS: m/z = 4679.8 [$Tb_2Gd_7(\mu$-OH$)_{10}$(Bu$)_{16}$]$^+$. Elemental analysis calculated for $C_{176}H_{218}NO_{61}Tb_2Gd_7$: C, 44.58%, H, 4.63%, N, 0.30%. Found: C, 44.32%, H, 4.54%, N, 0.26%.

**$Tb_5Gd_4$-Bu ([$Tb_5Gd_4(\mu$-OH$)_{10}$(Bu$)_{16}$]$NO_3$)**: Yield: 41%. Selected IR (ATR, $cm^{-1}$): 3573 (w, O–H), 3234 (w, O–H), 2956 (m, C–H), 1318 (s, C–O). FAB-MS: m/z = 4684.9 [$Tb_5Gd_4(\mu$-OH$)_{10}$(Bu$)_{16}$]$^+$. Elemental analysis calculated for $C_{176}H_{218}NO_{61}Tb_5Gd_4$: C, 44.53%, H, 4.63%, N, 0.30%. Found: C, 44.23%, H, 4.52%, N, 0.24%.

**$Tb_8Gd_1$-Bu ([$Tb_8Gd_1(\mu$-OH$)_{10}$(Bu$)_{16}$]$NO_3$)**: Yield: 49%. Selected IR (ATR, $cm^{-1}$): 3573 (w, O–H), 3234 (w, O–H), 2956 (m, C–H), 1318 (s, C–O). FAB-MS: m/z = 4689.8 [$Tb_8Gd_1(\mu$-OH$)_{10}$(Bu$)_{16}$]$^+$. Elemental analysis calculated for $C_{176}H_{218}NO_{61}Tb_8Gd_1$: C, 44.48%, H, 4.62%, N, 0.29%. Found: C, 44.24%, H, 4.53%, N, 0.24%.
**$Tb_9$-Bu ([$Tb_9(\mu$-OH$)_{10}$(Bu$)_{16}$]$NO_3$)**: Yield: 52%. Selected IR (ATR, $cm^{-1}$): 3573 (w, O–H), 3234 (w, O–H), 2956 (m, C–H), 1318 (s, C–O). FAB-MS: m/z = 4691.3 [$Tb_9(\mu$-OH$)_{10}$(Bu$)_{16}$]$^+$. Elemental analysis calculated for $C_{176}H_{218}NO_{61}Tb_9$: C, 44.47%, H, 4.62%, N, 0.29%. Found: C, 44.20%, H, 4.57%, N, 0.27%.

### *6.3.4 Spectroscopy*

Absorption spectra of $Tb_nGd_{9\text{-}n}$-Bu were obtained by using a JASCO V-670 spectrometer. Emission spectra were measured using a Horiba/Jobin-Yvon FluoroLog-sp3 spectrofluorometer and a JASCO FP-6600 spectrometer. The combination of an integration sphere and a JASCO FP-6600 spectrometer was used to measure emission quantum yields. Emission lifetimes were measured using the third harmonic (355 nm) of a Q-switched Nd:YAG laser. The temperature was controlled using an Oxford Instruments OptistatDN2 cryostat.

## 6.4 Results and Discussion

### *6.4.1 Structure and Identification*

The molecular structure and coordination geometry of the nine Ln(III) ions in $Tb_nGd_{9\text{-}n}$-Bu were elucidated by X-ray single-crystal analysis of $Gd_9$-Bu in Chap. 4 (see Sect. 4.3.1, Structure and Identification). Here, the structure is briefly reviewed (see Fig. 4.2a). The analysis showed that the Gd(III) ions in the cluster form an "hour-glass" structure in which the upper four and lower four Gd(III) ions are connected to the center Gd(III) ion via oxygen. All Gd(III) ions have an 8-coordination structure. The center Gd(III) ion is coordinated only by oxygen atoms and not by Bu ligands. Each of the Gd(III) ions in the outer unit is coordinated by both Bu ligands and oxygen atoms. The distances between two Gd(III) ions for all combinations are summarized in Table 6.4. From Continuous Shape Measure (CShM) [11, 12] calculations, Gd(III) ion in the center and Gd(III) ions in the outer unit was attributed to 8-square antiprism (8-SAP) and 8-trigonal dodecahedron (8-TDH) coordination geometry, respectively.

The structures of the other $Tb_nGd_{9\text{-}n}$-Bu ($n = 1, 2, 5, 8$) were compared by the combination of powder XRD (Fig. 6.5a) and FAB-MS (Fig. 6.5b). In the XRD results, the diffraction angle is corrected on the Si powder peak at $2\theta = 28.43°$ for precise comparison of the peaks. Focusing on the three distinguished peaks ($2\theta_1 = 11.3°$, $2\theta_2 = 15.6°$, and $2\theta_3 = 19.8°$), these peaks only showed shift within 0.02° indicating that the mixed Tb/Gd clusters have the same structure as those of $Tb_9$-Bu and $Gd_9$-Bu. This is further supported by FAB-MS spectroscopy which showed m/z value of the $Tb_nGd_{9-n}$-Bu corresponding to their calculated molecular weight without the $NO_3^-$ counter-anion. The results presented above indicate that all the clusters were successfully synthesized with the identical structure.

### 6.4.2 Photophysical Properties

The measured (normalized) emission spectra of $Tb_nGd_{9-n}$-Bu ($n = 1, 2, 5, 8, 9$) in $1.0 \times 10^{-4}$ M chloroform ($\lambda_{EX} = 380$ nm, ligand excitation) are shown in Fig. 6.6a. The shape of the Ln(III) emission spectrum is known to be influenced by the coordination geometry. The spectral shape of the emission was identical in all $Tb_nGd_{9-n}$-Bu indicating that the local geometry of the Tb(III) ions is retained in all of the clusters. Furthermore, the spectral shape of $Tb_9$-Bu in powder form (Fig. 6.6 inset) was also the same as that in the solution form, indicating that the coordination structure is maintained in solution [13–15]. The emission of $Tb_nGd_{9-n}$-Bu exhibited the characteristic 4f–4f transition of the Tb(III) ion with each of the peaks corresponding to its $^5D_4 \rightarrow {}^7F_J$ ($J = 6 - 1$) transitions [16]. The emission spectrum of $Gd_9$-Bu in $1.0 \times 10^{-4}$ M chloroform solution at 210 K (Fig. 6.6b) was measured to estimate the triplet excited state ($T_1$) energy level of the ligand. The excited state of Gd(III) ion is energetically higher than the excited states of the ligand, and thus the $T_1$ state is the lowest excited state. At sufficiently low temperature, the nonradiative process is suppressed, and the emission from the $T_1$ state was observed [17]. An emission peak was observed at 461 nm (21,690 $cm^{-1}$). Since the $^5D_4$ level of a Tb(III) ion

**Table 6.4** Distance between Gd(III) ions for all combinations (in Angstroms)

| 2nd Gd | Gd1 | Gd2 | Gd3 | Gd4 | Gd5 | Gd6 | Gd7 | Gd8 | Gd9 |
|---|---|---|---|---|---|---|---|---|---|
| Gd1 | – | 3.631 | 3.640 | 5.153 | 5.618 | 7.086 | 5.717 | 7.168 | 3.698 |
| Gd2 | 3.631 | – | 5.128 | 3.638 | 5.749 | 5.622 | 7.168 | 7.070 | 3.691 |
| Gd3 | 3.640 | 5.128 | – | 3.631 | 7.104 | 7.203 | 5.618 | 5.749 | 3.710 |
| Gd4 | 5.153 | 3.638 | 3.631 | – | 7.203 | 5.764 | 7.086 | 5.622 | 3.709 |
| Gd5 | 5.618 | 5.749 | 7.104 | 7.203 | – | 3.631 | 3.640 | 5.128 | 3.710 |
| Gd6 | 7086 | 5.622 | 7.203 | 5.764 | 3.631 | – | 5.153 | 3.638 | 3.709 |
| Gd7 | 5.717 | 7.168 | 5.618 | 7.086 | 3.640 | 5.153 | – | 3.631 | 3.698 |
| Gd8 | 7.168 | 7.070 | 5.749 | 5.622 | 5.128 | 3.638 | 3.631 | – | 3.691 |
| Gd9 | 3.698 | 3.691 | 3.710 | 3.709 | 3.710 | 3.709 | 3.698 | 3.691 | – |

The labels of the Gd ions in this table are based on the labels used in the Theoretical Background section and not the labels in the CIF file for consistency with the main text

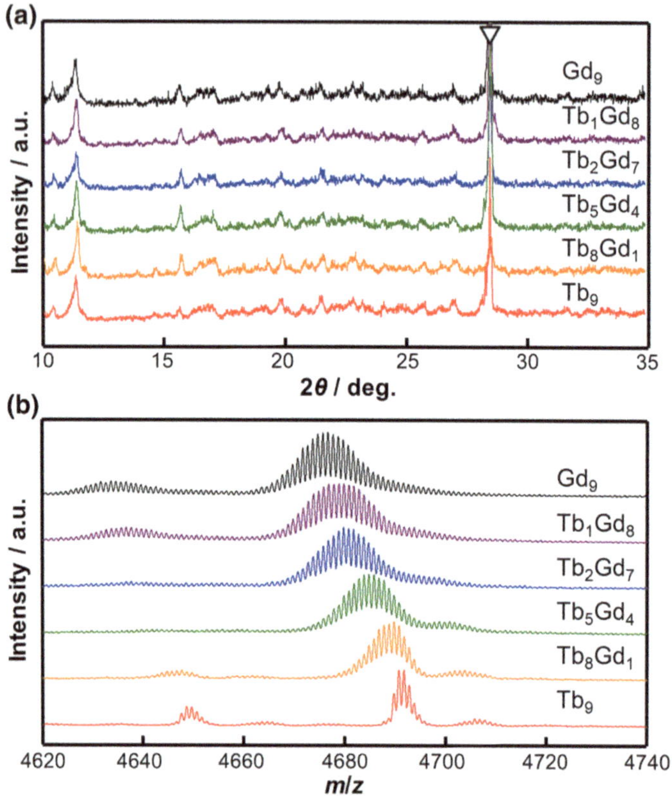


**Fig. 6.5** **a** Powder XRD pattern of $Tb_nGd_{9-n}$-Bu. The inverse triangle represents the peak of Si powder as a standard. **b** FAB-MS results of $Tb_nGd_{9-n}$-Bu. The broadening of the peak with increasing number of Gd(III) ions is due to the stable isotopes

is 20,620 $cm^{-1}$ as observed in the emission spectra (485 nm), the energy gap $\Delta E$ between $T_1$ and $^5D_4$ is 1070 $cm^{-1}$. This gap is well within the range for BET in Tb(III) complexes ($\Delta E$ <1850 $cm^{-1}$) and thus BET is expected to take place in $Tb_nGd_{9-n}$-Bu [18].

The emission lifetimes were determined by using a nanosecond pulse laser ($\lambda_{EX}$ = 355 nm, $\lambda_{EM}$ = 550 nm) for the clusters in a $1.0 \times 10^{-4}$ M chloroform solution. The emission decay profiles were single-exponential (Fig. 6.7a) in all the clusters. As summarized in Table 6.5, the lifetimes $\tau_{obs}$ increased with increasing number of Tb(III) ions in the $Tb_nGd_{9-n}$-Bu. The effect of TbET with increasing number of Tb(III) ions in the clusters was investigated by measuring the temperature dependence of emission lifetimes (Fig. 6.7b). A clear temperature dependence was observed above 270 K for $Tb_1Gd_8$-Bu and $Tb_2Gd_7$-Bu, while lifetimes remained fairly constant for $Tb_5Gd_4$-Bu, $Tb_8Gd_1$-Bu, and $Tb_9$-Bu. As mentioned in Chap. 5, the rate constant of BET $k_{BET}$ can be estimated by the following equation [19, 20]:

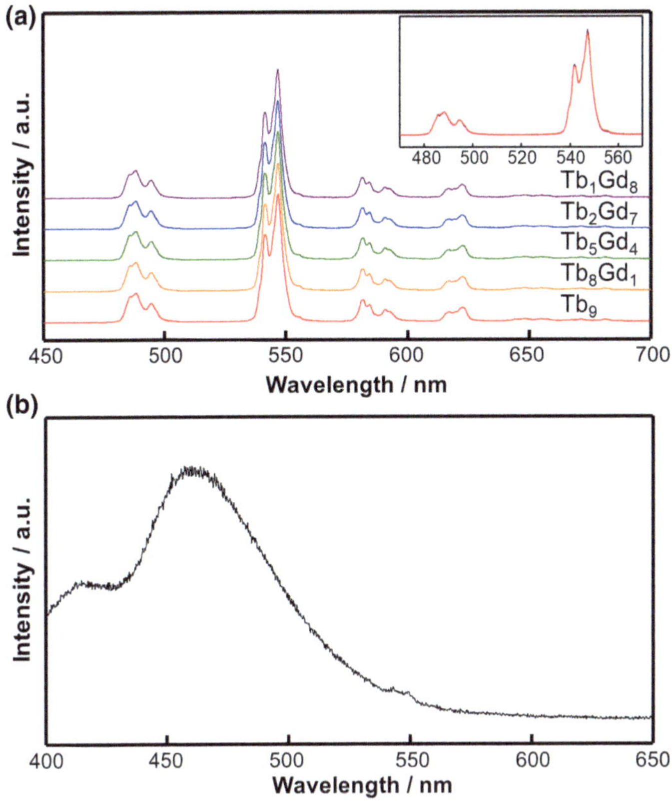


**Fig. 6.6** **a** Emission spectra of $Tb_nGd_{9-n}$-Bu (n = 1, 2, 5, 8, 9) in $1.0 \times 10^{-4}$ M chloroform solution ($\lambda_{EX} = 380$ nm). The spectra were normalized at the peak intensity and were applied an offset. Inset is the 470–570 nm region of the spectra ($^5D_4 \rightarrow {}^7F_6$ and $^7F_5$ transitions) of $Tb_1Gd_9$-Bu in solution (red) and powder (black). The spectra completely overlap. **b** Emission spectrum of $Gd_9$-Bu in $1.0 \times 10^{-4}$ M chloroform solution ($\lambda_{EX} = 380$ nm)

$$k_{BET} = \frac{1}{\tau(T)} - \frac{1}{\tau(87K)} = D\exp\left(-\frac{Ea}{k_B T}\right), \tag{6.12}$$

where $\tau$ is the measured emission lifetime at a given temperature, $\tau_{210K}$ is the emission lifetime at 210 K, $D$ is the frequency factor, $Ea_{BET}$ is the activation energy of BET, $k_B$ is the Boltzmann constant, and $T$ is the temperature. $Ea_{BET}$ and $k_{BET}$ can be calculated for $Tb_1Gd_8$-Bu by an Arrhenius plot within the range in which the temperature dependency was observed. These two values for the other clusters do not provide any meaningful result because other processes may play a factor in determining $k_{BET}$. The BET rate constant $k_{BET}$ and activation energy $Ea_{BET}$ for $Tb_1Gd_8$-Bu were calculated to be 167 $s^{-1}$ and 3235 $cm^{-1}$, respectively (Table 6.5).

Emission lifetime and its temperature dependence are in qualitative agreement with theoretical results presented in Sect. 6.2. The theoretical lifetime of $Tb_9$-Bu in

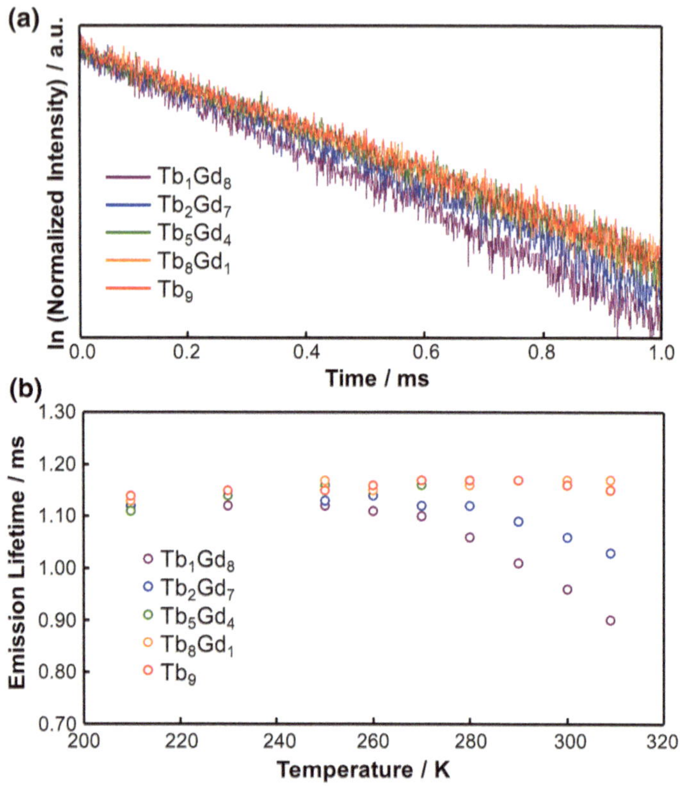


**Fig. 6.7** **a** Emission decay profiles of $Tb_nGd_{9-n}$-Bu (n = 1, 2, 5, 8, and 9) in $1.0 \times 10^{-4}$ M chloroform solution ($\lambda_{EX}$ = 355 nm, $\lambda_{EM}$ = 550 nm). **b** Plot of emission lifetime temperature dependence. The experimental uncertainties in the reported lifetimes are up to ±1.5%

the absence of BET was constant regardless of TbTbET rate $k_{TbTbET}$ (Table 6.6). As shown in Fig. 6.7b and Table 6.5, the experimental emission lifetimes of $Tb_nGd_{9-n}$-Bu at 210 K are the same value of approximately 1.12 ms. Meanwhile, the theoretical lifetime of $Tb_9$-Bu in the presence of BET shows that the lifetime is longer in the presence of TbTbET than in the absence of TbTbET. Indeed, experimental emission lifetimes at above 270 K in Fig. 6.7b clearly shows that the increase in number of Tb(III) ions (and thus faster TbTbET) leads to the increase in the emission lifetimes. Such agreement between the theoretical and experimental results qualitatively imply that the contribution of BET is suppressed in $Tb_9$-Bu because of the TbTbET and the existence of the center Tb(III) ion that is not in direct contact with the Bu ligands.

The obtained results provide new insight into a strategy to suppress BET in Ln(III) complexes. Formerly, the only strategy to suppress BET in Ln(III) complexes was to use ligands with high $T_1$ energy (>1850 $cm^{-1}$ in relative to the excited state of Ln(III) ion relevant to the FET process). This study now suggests that utilizing fast LnLnET in polynuclear Ln(III) complexes with at least one Ln(III) site that does not involve in the FET and BET process leads to suppression of BET without changing the $T_1$ energy of the ligands. Furthermore, since LnLnET is distance sensitive, Ln(III) clusters, in general, are a promising candidate as a material with low BET efficiency.

In terms of phonon-assisted energy transfer (discussed in Chap. 3), the bridging μ-OH ligands in the Ln(III) cluster core may be problematic due to high phonon energy mode. In this chapter, Tb(III) ions were used where the excited state energy is high enough for the PAETR to be a negligible process, but for lower excited state energy ions such as Yb(III), Er(III), and Nd(III) ions, the effect cannot be neglected. This means that the enhancement effect of LnLnET may be nullified due to PAETR in such low energy gap systems. However, more recently, chalcogenide-based Ln(III) clusters have been reported. Chalcogens (sulfur and selenium) are heavier atoms compared to oxygen and provides lower phonon-energy environment for the Ln(III) ions. Indeed, these clusters showed unprecedented high NIR luminescence efficiency (in the order of 10–30%), and may be suitable as the most efficiently luminescent functional material.

**Table 6.5** Photophysical properties of $Tb_nGd_{9-n}$-Bu

| Clusters | $\Phi_{\pi\pi*}$ [a] (%) | $\tau_{obs}$/ms [b] | $\tau_{210K}$/ms [b] | $D/10^8$ s$^{-1}$ [c] | $k_{BET}$/s$^{-1}$ [c] | $Ea_{BET}$/kJ mol$^{-1}$ [d] |
|---|---|---|---|---|---|---|
| $Tb_1Gd_8$ | 14 | 0.96 | 1.11 | 9.10 | 167 | 38.7 |
| $Tb_2Gd_7$ | 23 | 1.06 | 1.12 | * | * | * |
| $Tb_5Gd_4$ | 33 | 1.16 | 1.11 | * | * | * |
| $Tb_8Gd_1$ | 40 | 1.17 | 1.13 | * | * | * |
| $Tb_9$ | 39 | 1.16 | 1.14 | * | * | * |

Reproduced from Ref. [21] with permission from Springer Nature
[a]Measured in $1.0 \times 10^{-4}$ M chloroform solution ($\lambda_{EX} = 380$ nm)
[b]Measured in $1.0 \times 10^{-4}$ M chloroform solution ($\lambda_{EX} = 355$ nm)
[c]Calculated from Eq. (6.12). $D$ is the frequency factor
[d]Analyzed from an Arrhenius plot of Eq. (6.12) using lifetime temperature dependency results. $k_{BET}$ and $Ea_{BET}$ were only calculated for $Tb_1Gd_8$-Bu since other $Tb_nGd_{9-n}$-Bu ($n$ = 2, 5, 8, 9) involve TbTbET, which contributes to the temperature dependency of lifetimes

**Table 6.6** TbTbET rate constant dependency on the theoretical lifetime in the absence of BET

| $k_{TbET}$/s$^{-1}$ | $\tau_{calc}$/μs | $\Phi_{\pi\pi*,calc}$ (%) | $\eta_{BET,calc}$ (%) |
|---|---|---|---|
| 0 | 1300 | 25.28 | 0 |
| 500 | 1300 | 25.28 | 0 |
| 5000 | 1300 | 25.28 | 0 |
| 50000 | 1300 | 25.28 | 0 |

## 6.5 Conclusion

The role of ET between Tb(III) ions (TbTbET) in luminescent Tb(III) clusters in the reduction of BET has been demonstrated both theoretically and experimentally using $Tb_nGd_{9-n}$-Bu ($[Tb_nGd_{9-n}(\mu\text{-}OH)_{10}(Bu)_{16}]NO_3$) as a model system. In the Theoretical Motivation section, it was demonstrated that the two features of $Tb_9$-Bu were important for suppressing BET: (1) TbTbET by closely assembled Tb(III) ions and (2) the existence of center Tb(III) that is isolated from BET (and FET). Experimentally, the photophysical properties of nonanuclear $Tb_nGd_{9-n}$-Bu where $n$ = 0, 1, 2, 5, 8, and 9 were synthesized and investigated. The combination of X-ray single-crystal analysis, powder XRD, FAB-MS, and emission spectra revealed that these clusters have identical structures. Temperature dependence measurements of emission lifetime revealed that effect of BET becomes prominent in $Tb_nGd_{9-n}$-Bu at temperatures above 270 K. Below this temperature, the emission lifetime is constant for all clusters. Above 270 K, the decrease in emission lifetime is mitigated for clusters with over five Tb(III) ions. By comparing the trends observed in experimental results to those of theoretical results, it was found that the contribution of BET is indeed suppressed in clusters with a large number of Tb(III) ions. These findings provide new insight into the fundamental photophysics of Ln(III) complexes as well as an indication of a novel strategy to achieve higher luminescence efficiency [21].

## 6.6 Appendix

### *6.6.1 Dependence of TbTbET in Absence of BET*

TbTbET rate constant dependence on the lifetime of $Tb_9$-Bu in the absence of BET was evaluated by the same method presented in Sect. 6.2, but with different values for TbTbET rate constant $k_{TbTbET}$ (= 0, 500, 5000, and 50,000 $s^{-1}$). BET rate constant $k_{BET}$ is 0 $s^{-1}$, representing the absence of BET. Calculated lifetime $\tau_{calc}$, quantum yield $\Phi_{\pi\pi*,calc}$, and BET efficiency $\eta_{BET,calc}$ are summarized in Table 6.6. No dependence of TbTbET rate constant on the lifetimes and quantum yields were observed, unambigiously indicating that TbTbET serves no role in changing the photophysical properties of $Tb_9$-Bu in the absence of BET.

### 6.6.2 MATLAB Script for Theoretical Calculation of $Tb_9$-Bu

Theoretical calculations were performed using MATLAB software. The procedures are explained in Sect. 6.2. Below is an example of the script for the calculation of lifetimes and quantum yield.

Theoretical lifetime ($k_{TbTbET}$ = 0 $s^{-1}$, $k_{BET}$ = 3000 $s^{-1}$):

```
syms t s j krS1 knrS1 kisc krT1 knrT1 kPSEnT kBEnT krTbt knrTbt krTbs knrTbs kTbEnT
X1(t) X2(t) X3(t) X4(t) X5(t) X6(t) X7(t) X8(t) X9(t) X10(t) X11(t)

krS1 = 40000000
knrS1 = 340000000
kisc = 750000000
krT1 = 5000
knrT1 = 10000
kPSEnT = 6000
kBEnT = 3000
kTbEnT = 0
krTbt = 400
knrTbt = 400
krTbs = 400
knrTbs = 400

j = 1
I = eye(11)

C = [j; 0; 0; 0; 0; 0; 0; 0; 0; 0; 0]
A = [-(krS1+knrS1+kisc) 0 0 0 0 0 0 0 0 0 0;
   kisc -(krT1+knrT1+8*kPSEnT) kBEnT kBEnT kBEnT kBEnT kBEnT kBEnT kBEnT
kBEnT 0;
   0 kPSEnT -(krTbt+knrTbt+kBEnT+3.31*kTbEnT) kTbEnT kTbEnT 0.13*kTbEnT
0.07*kTbEnT 0.02*kTbEnT 0.07*kTbEnT 0.02*kTbEnT kTbEnT;
   0 kPSEnT kTbEnT -(krTbt+knrTbt+kBEnT+3.31*kTbEnT) 0.13*kTbEnT kTbEnT
0.07*kTbEnT 0.07*kTbEnT 0.02*kTbEnT 0.02*kTbEnT kTbEnT;
   0 kPSEnT kTbEnT 0.13*kTbEnT -(krTbt+knrTbt+kBEnT+3.31*kTbEnT) kTbEnT
0.02*kTbEnT 0.07*kTbEnT 0.02*kTbEnT 0.07*kTbEnT kTbEnT;
   0 kPSEnT 0.13*kTbEnT kTbEnT kTbEnT -(krTbt+knrTbt+kBEnT+3.31*kTbEnT)
0.02*kTbEnT 0.02*kTbEnT 0.07*kTbEnT 0.07*kTbEnT kTbEnT;
   0 kPSEnT 0.07*kTbEnT 0.07*kTbEnT 0.02*kTbEnT 0.02*kTbEnT -
(krTbt+knrTbt+kBEnT+3.31*kTbEnT) kTbEnT kTbEnT 0.13*kTbEnT kTbEnT;
   0 kPSEnT 0.02*kTbEnT 0.07*kTbEnT 0.07*kTbEnT 0.02*kTbEnT kTbEnT -
(krTbt+knrTbt+kBEnT+3.31*kTbEnT) 0.13*kTbEnT kTbEnT kTbEnT;
   0 kPSEnT 0.07*kTbEnT 0.02*kTbEnT 0.02*kTbEnT 0.07*kTbEnT kTbEnT 0.13*kTbEnT
-(krTbt+knrTbt+kBEnT+3.31*kTbEnT) kTbEnT kTbEnT;
```

```
0 kPSEnT 0.02*kTbEnT 0.02*kTbEnT 0.07*kTbEnT 0.07*kTbEnT 0.13*kTbEnT kTbEnT
kTbEnT -(krTbt+knrTbt+kBEnT+3.31*kTbEnT) kTbEnT;
0 0 kTbEnT kTbEnT kTbEnT kTbEnT kTbEnT kTbEnT kTbEnT kTbEnT -
(krTbs+knrTbs+8*kTbEnT)]
[P,D]=eig(A)
X = (P*expm(D*t)*P^-1)*C
X1(t) = X(1,1)
X2(t) = X(2,1)
X3(t) = X(3,1)
X4(t) = X(4,1)
X5(t) = X(5,1)
X6(t) = X(6,1)
X7(t) = X(7,1)
X8(t) = X(8,1)
X9(t) = X(9,1)
X10(t) = X(10,1)
X11(t) = X(11,1)
X22(t) = X2
X29(t) = X3+X4+X5+X6+X7+X8+X9+X10+X11

hold on
X29plot = ezplot(X29,[0,0.005])
set(X29plot, 'color','b','Linewidth',2)
X22plot = ezplot(X22,[0,0.005])
set(X22plot, 'Linestyle', '--', 'color','b','Linewidth',2)
Title = title('Tb9 Cluster','FontSize',15)
ylabel('Population Density','FontSize',10)
xlabel('Time / s','FontSize',10)
legend([X29plot X22plot],'All Tb','T1')
axis([0 0.005 0 1])

double(1/((log(X29(0.5))-log(X29(1)))/(1-0.5)))
```

Quantum yield ($k_{\mathrm{TbTbET}}$ = 50000 s$^{-1}$, $k_{\mathrm{BET}}$ = 3000 s$^{-1}$):

```
syms t s j krS1 knrS1 kisc krT1 knrT1 kPSEnT kBEnT krTbt knrTbt krTbs knrTbs kTbEnT
X1(t) X2(t) X3(t) X4(t) X5(t) X6(t) X7(t) X8(t) X9(t) X10(t) X11(t)

krS1 = 40000000
knrS1 = 340000000
kisc = 750000000
krT1 = 5000
knrT1 = 10000
kPSEnT = 6000
kBEnT = 3000
kTbEnT = 50000
krTbt = 400
knrTbt = 400
krTbs = 400
knrTbs = 400

j = 1
I = eye(11)

C = [j; 0; 0; 0; 0; 0; 0; 0; 0; 0; 0]
A = [-(krS1+knrS1+kisc) 0 0 0 0 0 0 0 0 0 0; kisc -(krT1+knrT1+8*kPSEnT) kBEnT kBEnT
kBEnT kBEnT kBEnT kBEnT kBEnT kBEnT 0; 0 kPSEnT -
```

```
(krTbt+knrTbt+kBEnT+3.31*kTbEnT) kTbEnT kTbEnT 0.13*kTbEnT 0.07*kTbEnT
0.02*kTbEnT 0.07*kTbEnT 0.02*kTbEnT kTbEnT; 0 kPSEnT kTbEnT -
(krTbt+knrTbt+kBEnT+3.31*kTbEnT) 0.13*kTbEnT kTbEnT 0.07*kTbEnT 0.07*kTbEnT
0.02*kTbEnT 0.02*kTbEnT kTbEnT; 0 kPSEnT kTbEnT 0.13*kTbEnT -
(krTbt+knrTbt+kBEnT+3.31*kTbEnT) kTbEnT 0.02*kTbEnT 0.07*kTbEnT 0.02*kTbEnT
0.07*kTbEnT kTbEnT; 0 kPSEnT 0.13*kTbEnT kTbEnT kTbEnT -
(krTbt+knrTbt+kBEnT+3.31*kTbEnT) 0.02*kTbEnT 0.02*kTbEnT 0.07*kTbEnT
0.07*kTbEnT kTbEnT; 0 kPSEnT 0.07*kTbEnT 0.07*kTbEnT 0.02*kTbEnT 0.02*kTbEnT -
(krTbt+knrTbt+kBEnT+3.31*kTbEnT) kTbEnT kTbEnT 0.13*kTbEnT kTbEnT; 0 kPSEnT
0.02*kTbEnT 0.07*kTbEnT 0.07*kTbEnT 0.02*kTbEnT kTbEnT -
(krTbt+knrTbt+kBEnT+3.31*kTbEnT) 0.13*kTbEnT kTbEnT kTbEnT; 0 kPSEnT
0.07*kTbEnT 0.02*kTbEnT 0.02*kTbEnT 0.07*kTbEnT kTbEnT 0.13*kTbEnT -
(krTbt+knrTbt+kBEnT+3.31*kTbEnT) kTbEnT kTbEnT; 0 kPSEnT 0.02*kTbEnT
0.02*kTbEnT 0.07*kTbEnT 0.07*kTbEnT 0.13*kTbEnT kTbEnT kTbEnT -
(krTbt+knrTbt+kBEnT+3.31*kTbEnT) kTbEnT; 0 0 kTbEnT kTbEnT kTbEnT kTbEnT
kTbEnT kTbEnT kTbEnT kTbEnT -(krTbs+knrTbs+8*kTbEnT)]
[P,D]=eig(A)
X = (A^-1)*((P*expm(D*t)*P^-1)-I)*C
X1(t) = X(1,1)
X2(t) = X(2,1)
X3(t) = X(3,1)
X4(t) = X(4,1)
X5(t) = X(5,1)
X6(t) = X(6,1)
X7(t) = X(7,1)
X8(t) = X(8,1)
X9(t) = X(9,1)
X10(t) = X(10,1)
X11(t) = X(11,1)

Xtbqy1(t) = krTbt*(X3+X4+X5+X6+X7+X8+X9+X10)+krTbs*X11
double(Xtbqy1(inf))

Xalltbdecay(t) = (krTbt+knrTbt)*(X3+X4+X5+X6+X7+X8+X9+X10)+(krTbs+knrTbs)*X11
double(Xalltbdecay(inf))
```

# References


1. van der Ende BM, Aarts L, Meijerink A (2009) Lanthanide ions as spectral converters for solar cells. Phys Chem Chem Phys 11:11081–11095
2. Huang X, Han S, Huang W, Liu X (2013) Enhancing solar cell efficiency: the search for luminescent materials as spectral converters. Chem Soc Rev 42:173–201
3. Shang Y, Hao S, Yang C, Chen G (2015) Enhancing solar cell efficiency using photon upconversion materials. Nanomaterials 5:1782–1809
4. Gautier R, Li X, Xia Z, Massuyeau F (2017) Two-step design of a single-doped white phosphor with high color rendering. J Am Chem Soc 139:1436–1439
5. Starzak ME (1989) Mathematical methods in chemistry and physics. Springer
6. Malta OL (2008) Mechanisms of non-radiative energy transfer involving lanthanide ions revisited. J Non Cryst Solids 354:4770–4776
7. Kushida T (1973) Energy transfer and cooperative optical transitions in rare-earth doped inorganic materials. II. Comparison with experiments. J Phys Soc Jpn 34:1327–1333
8. Aboshyan-Sorgho L, Nozary H, Aebischer A, Bünzli J-CG, Morgantini PY, Kittilstved KR, Hauser A, Eliseeva SV, Petoud S, Piguet C (2012) Optimizing millisecond time scale near-

infrared emission in polynuclear chrome (III)–lanthanide (III) complexes. J Am Chem Soc 134:12675–12684
9. Tobita S, Arakawa M, Tanaka I (1985) The paramagnetic metal effect on the ligand localized S1. apprx.fwdarw. T1 intersystem crossing in the rare-earth-metal complexes with methyl salicylate. J Phys Chem 89:5649–5654
10. Yang Y, Li J, Liu X, Zhang S, Driesen K, Nockemann P, Binnemans K (2008) Listening to lanthanide complexes: determination of the intrinsic luminescence quantum yield by nonradiative relaxation. ChemPhysChem 9:600–606
11. Casanova D, Llunell M, Alemany P, Alvarez S (2005) The rich stereochemistry of eight-vertex polyhedra: a continuous shape measures study. Chem Eur J 11:1479–1494
12. Pinsky M, Avnir D (1998) Continuous symmetry measures. 5. The classical polyhedra. Inorg Chem 37:5575–5582
13. Lima NBD, Gonçalves SMC, Júnior SA, Simas AM (2013) A comprehensive strategy to boost the quantum yield of luminescence of europium complexes. Sci Rep 3:2395
14. Miyata K, Nakagawa T, Kawakami R, Kita Y, Sugimoto K, Nakashima T, Harada T, Kawai T, Hasegawa Y (2011) Remarkable luminescence properties of lanthanide complexes with asymmetric dodecahedron structures. Chem Eur J 17:521–528
15. Yanagisawa K, Nakanishi T, Kitagawa Y, Seki T, Akama T, Kobayashi M, Taketsugu T, Ito H, Fushimi K, Hasegawa Y (2015) Seven-coordinate luminophores: brilliant luminescence of lanthanide complexes with C3v geometrical structures. Eur J Inorg Chem 2015:4769–4774
16. Beitz JV (1994) f-State luminescence of trivalent lanthanide and actinide ions in solution. J Alloys Compd 207–208:41–50
17. Bassett AP, Magennis SW, Glover PB, Lewis DJ, Spencer N, Parsons S, Williams RM, De Cola L, Pikramenou Z (2004) Highly luminescent, triple-and quadruple-stranded, dinuclear Eu, Nd, and Sm (III) lanthanide complexes based on bis-diketonate ligands. J Am Chem Soc 126:9413–9424
18. Latva M, Takalo H, Mukkala V-M, Matachescu C, Rodríguez-Ubis JC, Kankare J (1997) Correlation between the lowest triplet state energy level of the ligand and lanthanide (III) luminescence quantum yield. J Lumin 75:149–169
19. Miyata K, Konno Y, Nakanishi T, Kobayashi A, Kato M, Fushimi K, Hasegawa Y (2013) Chameleon luminophore for sensing temperatures: control of metal-to-metal and energy back transfer in lanthanide coordination polymers. Angew Chem Int Ed 52:6413–6416
20. Yanagisawa K, Kitagawa Y, Nakanishi T, Akama T, Kobayashi M, Seki T, Fushimi K, Ito H, Taketsugu T, Hasegawa Y (2017) Enhanced luminescence of asymmetrical seven-coordinate $Eu^{III}$ complexes including LMCT perturbation. Eur J Inorg Chem 2017:3843–3848
21. Omagari S, Nakanishi T, Kitagawa Y, Seki T, Fushimi K, Ito H, Meijerink A, Hasegawa Y (2016) Critical role of energy transfer between terbium ions for suppression of back energy transfer in nonanuclear terbium clusters. Sci Rep 6:37008

# Chapter 7
# Summary and Outlook

**Abstract** Throughout this thesis, various aspects of energy transfer processes in polynuclear lanthanide complexes have been explored. The purpose of this chapter is to present the conclusion of each of the chapters and to summarize them so that the core concept of the thesis can be discussed and concluded. An outlook based on the conclusions of this thesis is also provided.

## 7.1 Summary and Conclusion

Polynuclear Ln(III) complexes gained considerable attention recently for their ability to utilize interaction between Ln(III) ions, and thus their potential as functional luminescent material may exceed those of mononuclear complexes. The objective of this study was to unravel various factors in polynuclear Ln(III) complexes (namely coordination polymers and clusters) that affect their luminescence efficiency, with emphasis on energy transfer processes. Such factors include concentration quenching, spin-orbit coupling, and back energy transfer. Uncovering these factors could help understand luminescence mechanism of Ln(III) complexes in general and also allow for more rational design of Ln(III) complexes.

Chapter 1 was a background on luminescence in general and lanthanide complexes and introduced the concepts that are relevant to the studies shown in this thesis. The photophysical processes in general, particularly radiative and vibrational relaxation, were explained with the least amount of mathematical formulations possible and with more conceptual figures. Then the photophysical properties of lanthanides and Ln(III) complexes were explored along with the differences between them and typical phosphors. Finally, polynuclear Ln(III) complexes are introduced with several examples.

In Chapter 2, the theory behind lanthanide luminescence as well as kinetic analysis of complicated systems was explained. The former reviews the Judd-Ofelt theory as well as the theory of energy transfer under the framework of the Judd-Ofelt theory and explains the derivation of the rate constant of luminescence and energy transfer. The latter shows the derivation of rate equations for a system of complicated excited state dynamics under an impulse and a constant excitation source.



S. Omagari, *Energy Transfer Processes in Polynuclear Lanthanide Complexes*, Springer Theses, https://doi.org/10.1007/978-981-13-6049-7_7

Several assumptions in the calculations in the latter chapters were based on the implication of the theories described in this chapter.

The consequence of energy transfer between Ln(III) ions is the concentration quenching, and their effects in Yb(III) coordination polymer was described in Chap. 3. The emission spectra and lifetimes of Yb(III) emission of the polymer were recorded and compared to the theoretical lifetimes under different concentration quenching model. These models are no-concentration quenching, killer-site, and phonon-assisted energy transfer relaxation. The theories described in Chap. 2 was extensively used for the derivation of theoretical lifetimes. The experimental results were qualitatively consistent with the phonon-assisted energy transfer relaxation model, indicating that this is the concentration quenching mechanism in Yb(III) coordination polymers. Furthermore, the quantitative aspects were also investigated, indicating the validity of the model.

The central discussion described in Chap. 4 was the effect of spin-orbit coupling on the sensitization efficiency of Ln(III) complexes. Specifically, the sensitization efficiency of Yb(III)/Gd(III) or Yb(III)/Lu(III) mixed nonanuclear Ln(III) clusters was discussed. Yb(III) ion is the emitting center, and Gd(III) and Lu(III) ions (both non-emissive) were used to control the spin-orbit coupling strength. The sensitization efficiency is dependent on the spin-orbit coupling that enhances both the $S_1 \rightarrow T_1$ intersystem crossing and the $T_1 \rightarrow S_0$ relaxation, which the former and the latter affects positively and negatively, respectively, to the sensitization efficiency. The emission spectra of the clusters indicated that strong spin-orbit coupling did not necessarily lead to high sensitization efficiency because it promoted $S_1 \rightarrow T_1$ intersystem crossing but also the $T_1 \rightarrow S_0$ relaxation. This is contrary to what was the usual convention that strong spin-orbit coupling is always good for high sensitization efficiency, and the conclusion presented in this chapter is therefore an important insight into photophysics of Ln(III) complexes.

In Chap. 5, the factors contributing to the back energy transfer rate constant in nonanuclear Tb(III) clusters were reported. The chapter also reported the effect of intramolecular interaction between alkyl salicylate ligands on the photophysical properties of the clusters. The result shows that the core structure of the cluster is rigid, and maintains the luminescent property of Tb(III) ions that are sensitive to the coordination environment. Provided with these results, methyl salicylates with methyl substituent on the aromatic part were used as ligands with modified excited state energy levels. Temperature dependent emission lifetime measurement revealed that back energy transfer is not only dependent on the energy gaps between $T_1$ and $^5D_4$ state of Tb(III) ion but also the activation energies and frequency factors. These two factors are of great importance in terms of the fundamentals of energy transfer between ligands and Ln(III) ions, where the frequency factors mean the electronic correlation between two states (ligand excited state and Ln(III) excited state) and the activation energy mean the structural changes during energy transfer.

Finally, in Chap. 6, the effect of the energy transfer between Tb(III) ions on the back energy transfer was described. The theoretical motivation section explained using kinetic analysis, the possibility of fast energy transfer between Tb(III) ions to suppress the contribution of back energy transfer in nonanuclear Tb(III) clusters by delivering the energy to a Tb(III) ion site that is not involved in forward nor back energy transfers. Experimental results, particularly the temperature dependence of emission lifetimes of Tb/Gd mixed clusters, showed qualitatively similar trend with the theoretical lifetimes, indicating that back energy transfer is indeed suppressed by energy transfer between Tb(III) ions. This is a radically new approach in suppressing back energy transfer, which formerly was considered only to be suppressed by raising the $T_1$ energy of the ligands.

The commonly reported strategies for Ln(III) complexes in general are to distort the ligand field of a Ln(III) ion to raise radiative rate, use of low phonon energy modes in the ligands to suppress vibrational relaxation, and optimize $T_1$ energy to improve sensitization efficiency. Of course, these strategies can also be applied for polynuclear Ln(III) complexes. This thesis, on the other hand, focused on new strategies centered around polynuclear Ln(III) complexes, a new class of luminescent material with potential functionally that may go beyond the ability of mononuclear counterparts. Herein, as a summary of the thesis, the author present three novel strategies for improving luminescence efficiency in polynuclear Ln(III) complexes:

(1) High vibronic modes are eliminated to suppress phonon-assisted energy transfer relaxation.
(2) Spin-orbit coupling strength is kept optimum so that $T_1 \rightarrow S_0$ relaxation will not overcome the enhancing effect of $S_1 \rightarrow T_1$ intersystem crossing have on sensitization efficiency.
(3) Fast energy transfer proceeds between Ln(III) ions to deliver the obtained energy to a more efficiently emitting Ln(III) site to counter a quenching process.

The last two strategies require that the Ln(III) ions are closely assembled. In the second strategy, spin-orbit coupling can be controlled by mixing different Ln(III) ions but is only possible if the Ln(III) ions are close enough to the ligands. For the third strategy, fast energy transfer between Ln(III) ions means close distance between Ln(III) ions. Ln(III) clusters are the only type of Ln(III) complexes that can satisfy these criteria. The first strategy is somewhat difficult for Ln(III) clusters since most them contain bridging –OH ligands, which has high energy vibration (O–H vibration). This means that not only the phonon-assisted energy transfer relaxation is enhanced, but also typically the nonadiabatic vibrational relaxation. Fortunately, recent advances in coordination chemistry allowed chalcogenide-based Ln(III) clusters that exhibit very efficient luminescence even for near infrared emitting lanthanides due to the heavy chalcogen elements that suppresses vibration. Therefore, the three strategies can be utilized for Ln(III) clusters, which indicates the promising properties Ln(III) clusters over other form of polynuclear Ln(III) complexes.

## 7.2 Outlook

With "sustainability" increasingly becoming the prerequisite of technological development, raising the efficiency of any device is now just as important as developing a new one. That means using more abundant and cheaper alternative to currently used material, and if that is not available then maximizing the efficiency. Although rare-earth elements are still relatively abundant in Earth's crust and a rare-earth crisis is likely to be a problem of much later generations, it does not mean that rare-earth is never going to be depleted. Therefore, it is perfectly reasonable to consider the maximization of the luminescence efficiency of lanthanide-based material to be one of the challenges today. The author believes that this thesis contributes to the effort in tackling such challenge by presenting radically new approaches.

The study described in this thesis provides insights into the fundamental properties not only of Ln(III) clusters but of Ln(III) complexes in general, the advantages of using Ln(III) clusters as efficiently luminescent functional material, and novel strategies to improve the luminescence efficiency of Ln(III) complexes. The strategies includes energy transfer between Ln(III) ions, spin-orbit coupling, the photophysical details of back energy transfer, and utilization of complicated dynamics. These concepts are radically different compared to the previous strategies, i.e. suppression of radiative and nonradiative processes of Ln(III) ions and optimization of excited state energy of ligands. Therefore, the author believes that the study shown in this thesis sets the stage for further research in the new quantum yield improvement strategies.

The numerous potential applications of Ln(III) complexes include, but not limited to, lighting, OLEDs, lasers, bioimaging, sensors, and spectral converters. However, the use of Ln(III) complexes is somewhat still limited despite their advantages. The author hopes that the introduction of the strategies and the insights provided in this thesis help pave a way to a breakthrough in technology and society.

# Curriculum Vitae

**Shun Omagari**
Department of Materials Science and Engineering, Tokyo Institute of Technology
S8-607, 2-12-1 Ookayama, Meguro, Tokyo 152-8552, Japan
E-mail: omagari.s.aa@m.titech.ac.jp
Web: http://www.op.titech.ac.jp/lab/vacha/.

**Appointments**
JSPS Research Fellowship for Young Scientists (PD), Tokyo Institute of Technology (Apr. 2018–).
JSPS Research Fellowship for Young Scientists (DC1), Hokkaido University (Apr. 2016–Mar. 2018).

**Education**
Doctor of Philosophy (Ph.D.) in Engineering, Hokkaido University (Oct. 2015–Mar. 2018).
Master of Chemical Sciences and Engineering, Hokkaido University (Apr. 2014–Sep. 2015).
Bachelor of Engineering, Hokkaido University (Apr. 2010–Mar. 2014).

**Grants**
Murata Science Foundation (2015).

**Awards**
Presentation Award (67th Conference of Japan Society of Coordination Chemistry).
Presentation Award (64th JSAP Spring Meeting).
Presentation Award (Annual Meeting on Photochemistry 2016).
Student Poster Award (Pacifichem 2015).



S. Omagari, *Energy Transfer Processes in Polynuclear Lanthanide Complexes*, Springer Theses, https://doi.org/10.1007/978-981-13-6049-7

Presentation Award (2015 Summer Research Meeting of Hokkaido Branch of the Chemical Society of Japan).
Presentation Award (2015 Winter Research Meeting of the Hokkaido Branch of Joint Chemistry Societies).
Poster Award (3rd Seminar Meeting of 2014 Hokkaido Branch of Young Coordination Chemist's Association).
Poster Award (Annual Meeting on Photochemistry 2014).
Poster Award (26th Meeting on Photochemistry of Coordination Compounds).

MIX
Papier aus verantwortungsvollen Quellen
Paper from responsible sources
FSC® C105338

If you have any concerns about our products,
you can contact us on
**ProductSafety@springernature.com**

In case Publisher is established outside the EU,
the EU authorized representative is:
**Springer Nature Customer Service Center GmbH**
**Europaplatz 3, 69115 Heidelberg, Germany**

Printed by Libri Plureos GmbH
in Hamburg, Germany